THE SINGULARITY IS COMING
(The Artificial Intelligence Explosion)

But the AGI robots will be here first!

THE SINGULARITY IS COMING
(The Artificial Intelligence Explosion)
But the AGI robots will be here first

Tony Thorne MBE
© 2014-7 Tony Thorne MBE (Sections 1 and 2)
This is update number 4, March 2019

First ASIN: **B00N1RJ0L4**

A Chinese language version of the first edition of this book was published in 2016 by P.T. Press, The Chinese Government IT Division, in Beijing

Author's Note: Smart computers and machines are already with us. Next will come intelligent AGI robots, new mobile life forms far smarter than we are, followed by the inevitable Singularity. Is this the beginning of a Golden Age for us all as optimistic experts predict... or could it be a total disaster for humanity, as other experts like Professor Stephen Hawking and Elon Musk have warned? Will our new intelligent creations become our slaves or our masters? After you have finished reading what follows you can decide for yourself, but don't expect it to be easy.

This revised eBook has been updated with new information for 2019, but first published in August 2014 by Etcetera Press.

Publisher's Note: AI (Artificial Intelligence) is an extremely important subject that affects us all, but its eventual outcome is uncertain. This particular 'scrapbook type compilation' is meant for general readers of all ages and explains the situation to date. It is also "ongoing" and may be revised with relevant new material, by the author as convenient. His website is constantly updated with news about these developments. (www.tonythornembe.com)

Editor's Introduction to the first edition: "The author of the Personal Text Sections of this book has a rare gift. He is a very see thinker, and it comes across in his writing. This is a rare talent, so I appreciate him letting me review his work." – Autumn Conley, USA, Professional Editor and Reviewer.

Acknowledgements: Other edited comments and sections in this compilation, not from the author/compiler, owe thanks to Wikipedia, several international English language newspapers, the BBC, Ray Kurzweil's AI Newsletter, Discovery Magazine, Singularity Hub (http://singularityhub.com/) and the AI forums etc. on Linked-In and several other websites. Plus, encouragement and help from my publishers, in the USA and China and other sources far too numerous to mention.

Section Contents

Futurists and Engineers, i.e. the people in the know, for and against, including comments by the author/compiler of this book.

4. **Page 105. Further Reading List**: A selection with extracts, notes and some reviews.

5. **Page 120. About the Author/Compiler** with some reviews of a few of his Books, including this one.

Section 1 - THE SINGULARITY IS COMING
What is a singularity?
Briefly for now, a singularity is best described as a one-way event, something that cannot be reversed once it occurs. It seems the term first came about in reference to mass being inevitably drawn into a black hole once its event horizon is approached. For mass, this means planets, stars, gas clouds, anything that gets in the way, even small galaxies, all consumed.

An event horizon is the point at which the gravitational force is so strong that even light cannot escape from it. Thus, we can never, see a black hole, yet its external effects can be observed and measured. For those readers who are not familiar with the concept of a black hole, it's a massive star that has collapsed; it becomes so compressed that it has a gravity force field strong enough to prevent light photons from escaping it. In other words, its escape velocity is greater than the speed of light, so it is invisible.

Escape velocity is the minimum speed required to completely escape from the pull of any particular gravitational field. Planet Earth's escape velocity is only around 25,000 miles per hour; whereas light travels at just over 186,000 miles in only one second.

Now let's discuss THE SINGULARITY. The awesome event that is coming closer every day. Perhaps you have not heard much about it from the media yet, but you soon will. Many scientists, engineers, and other experts think it poses a far greater threat to humanity than global warming or even nuclear war. They seem convinced it may begin in as few as ten years from the time this book was started; that means it could happen as early as 2024!

Many futurists and even a few authorities believe some factors will begin even earlier than that, and they forecast at least a general sixty percent unemployment situation in the developed nations very soon now, with even more to follow. When that occurs, the developed world's economic situation will change, dramatically... it must!

The particular aspects of the singularity this book is about are the developments that will inevitably follow on after the smart machines we already have today, beginning with artificial general intelligence (AGI), then artificial super intelligence (ASI) not long after. The remarkable smart developments that are part of our everyday lives now are just precursors to what we will experience in the near future. We can envision smarter smart-phones and even better iPads and tablets, because our clever gadgets and other devices are doing more and getting smarter every time they are updated automatically via the internet. Many of them already know even more than everything you know about, or even don't know about at the moment!

This book is about the incredibly clever developments that will come after that, then after that, and so on, beginning with artificial general intelligence (AGI), then inevitably artificial super intelligence (ASI). The latter will be THE SINGULARITY. That event will arrive soon after the first intelligent software algorithms produce AGI, after being downloaded into an assembly of parallel computer processing and memory chips large enough to contain them. This unprecedented event will be mind-blowing, and the inevitable known and unknown consequences, good or bad, will change our way of life completely. We'd better be ready for it... or else.

Before you read too deeply into this book, perhaps a little about my relevant background is in order. My first encounter with the

concept of a humanoid robot took place in my youth, some time ago now, when I first viewed Fritz Lang's silent film, Metropolis. Yes, it was that long ago..!
(http://www.imdb.com/title/tt0017136/)

Not long after that, I read a copy of the play, R.U.R. by Karel Capek. He was the first person to use the word "robot"; however, I soon realized that his story was really about androids, artificial biological creations, rather than mechanical robots we think of now.

(http://www.umich.edu/~engb415/literature/pontee/RU R/RURsmry.html)

After that came the AGI robot, **Robby**, in that ground-breaking 1956 film, Forbidden Planet, loosely based on Shakespeare's play, The Tempest, which carried a warning message now very relevant to our future. i.e. How the consequences of an advanced technology can wipe out a civilization. If you haven't seen it... you should.

 http://en.wikipedia.org/wiki/Forbidden_Planet)

Soon after that, one happy day, I came across a used copy of the American sci-fi magazine, Thrilling Wonder Stories. What a find that was for me back then, opening my eyes and my imagination to robots, aliens, and future science, a real world of wonder!
(http://en.wikipedia.org/wiki/Wonder_Stories)

I was inspired to write my first published story soon after that and was even paid for it. It was about the awesome perils of time travel and had nothing to do with robots, those tales came later. Shortly after, at my first SF Fan Convention in London, I met and befriended several other speculative fiction authors, but that's another story, and quite a list.

Many years later, in the late seventies, I bought one of the first home computers available in the island town where I lived at the time. Once I mastered that Apple II, I bought several more and set up a training school that became quite successful. After that, with a basic knowledge of programming techniques, I became involved in the development of semi-intelligent code-writing software, the first being the C.O.R.P. system (Combined, Operating, Re-entrant Programming).

Later, I moved on to the much more advanced Enterprise System. Both of these suites of interacting programs were initially created by American software genius, Alex Maromaty. I bought the rights to the latter system from him and set about developing it further, with many more useful features and additions, such as color text and screens. Alex and I called these systems, Program (code) Generators.

After the user answered and entered questions about what requirements were wanted, both programs were capable of generating the necessary algorithms, yes the software code itself, by themselves. This is the essential ability required by all artificially intelligent software, in order for it to function and learn as intended. An algorithm that can write code, then run it to see what it does, learn from it, and then write improved code by itself and run it again and so on, is now called **learning software.** It's what very SMART devices use, and nowadays there's quite a lot of it about... and more will follow.

In other words, an algorithm can be designed to analyze what it generates each time it's run, then make any required changes to itself, run itself again, and then continue the process as required. Of course, all this continues at lightning speed, inside the computer to which it has been downloaded, apart from the pauses it may make to ask for any more information it needs to complete the program... ready to run

I successfully developed all kinds of business software programs from scratch with the Enterprise System. My first, a double-entry book-keeping program was quite successful and sold reasonably well. I recall I also wrote an algorithm to control a simple mobile robot that was used at exhibitions as a crowd puller. It worked fine, but that robot was infinitely far from being intelligent and was connected to its controlling computer via a cable.

Getting back to the coming artificial intelligence explosion, let us consider how we human beings go about doing the things we do every day, in addition to being aware of our own existence, the functions an AGI robot, by definition, will also be able to undertake.

Apparently, each human brain has about 100 billion neurons (cells that communicate with each other) with about 100 trillion connections, each of which can perform computations of all kinds at a rate of about 200 per second. This approximates to 20 million billion brain "software" computation processes per second, a remarkably large number that evolution has developed for us over the many thousands of years of our existence. It also represents a much larger amount of computing power than even the most advanced laptops available have today, at the time of writing.

That said, some scientists have predicted that possibly only ten years from the time of the writing of this book, an artificial computer brain will be assembled. It will be able to achieve about half of the above-mentioned human brain computing power, based on the way things are now progressing. However, according to what's known as Moore's Law, only thirteen or fourteen months after that, an updated computer brain hardware will achieve power equality with our brains. If the necessary intelligent software is also ready to be downloaded into it as expected, then AGI will have arrived.

This is a sobering thought, and if the result also automatically produces consciousness (i.e. self-awareness and free-will), as some scientists believe, then a **new intelligent life form** will have been created. It will be able to work 24 hours a day, 7 days a week for no pay, easily doing the work that 2 to 3 humans did before. Think about what that situation will mean to an employer.

The United Nations might have to start thinking about the need to debate, and introduce, a non-human rights charter! This all might sound like the sci-fi I've always been a fan of, but there's really nothing impossible about the idea of a new intelligent life form being created! Also, given access to a suitable computer automated manufacturing plant, containing the latest developments in nano-technology and 3D printing, it will also be capable of continually replicating and improving itself. That ominous thought worries a lot of enlightened people.

We can surmise from the pace of modern developments that the hardware, the brain-like architecture, may be ready in a decade or less, but what about the intelligent software that will bring it to life? Will that be complete and ready as well, and how comprehensive will it be? Most importantly, will it be safe? These are all vital considerations for our future, but let's take a look at what's going been going on recently.

Two very smart computer brains have already been demonstrated, much to the amazement of even the most critical sceptics. These were the IBM Big Blue and Watson machines. The former could probably beat any human anywhere at a game of chess; even world champion chess player Gary Kasparov lost to it twice not so long ago. The Watson machine defeated even the most intelligent human contestants in the tricky TV game, Jeopardy, in which the competitors have to work out the questions to given answers, a far, far more complicated exercise than choosing the correct multiple-choice answer in those million-dollar TV shows.

But do those two machines really think? Well, in the case of Big Blue, before every move, it looked up every significantly relevant response, then employed the best move to find the next, and then the next, and so on, until it reached the most optimal end result. Big Blue accomplished this process much faster than any human chess player. Watson however, looked up every possible fact about the given answer, then researched that answer again and again, narrowing things down until it generated the most suitable question for the answer. All of this completed at an incredibly high speed that no human competitor's brain could possibly mimic.

The third major advance in intelligent (self-learning) software i.e. a remarkable program named AlphaGo, was created by scientists in Google's Deep Mind Division, and about ten years earlier than anyone thought possible. This latest machine appears to use the same parallel neural network processing that the human mind performs, but more accurately, much faster, and with instant access to a vast database. There's more about it in a later section of this book.

Faced with these developments, the sceptics prefer to redefine *"Thinking"*. **They limit it to being something only humans can do.** Then they continue with this claim no matter how many smarter developments are made. However, even they must soon realize that all the things only humans can do are becoming fewer and fewer every day. Then what?

About every year and a half, an interval that is also reducing, the power of computer chips doubles. They are also shrinking in size; hence, they require less energy to work even faster. The famous Moore's Law holds true, and this is expected to continue until chips reach down to the size of molecules and their processing speeds approach that of light itself. There's still quite a way to go, but many scientists reckon that in less than another eight years, the first computer brain will be created equal in 'computing power' to ours. Progress is exponential and not linear.

It can be surprisingly difficult for us to appreciate the outcome of a process that doubles each time, but we can put it in terms of this imaginary example to help us get a better grasp of the concept. Consider folding over a large, imaginary sheet of paper. When you do, the resulting thickness will be double that of the original. If you do this again and again and again, and so on, in less folds than you might imagine, the resulting theoretical thickness will eventually reach about a quarter of the way to the moon. The next 'fold' will get you halfway to it and after that, just one more 'fold' will get you all the way there! Now think about the 'folds' after that! In this same way, technology is moving in seemingly impossible, ever larger, leaps and bounds.

Computer brain power equality will be achieved very quickly in the last couple years of this cycle. Three years before, it will only have about one-eighth of the power required, from only one-sixteenth of it the year before that. In that last remarkable and historical year, it will double again and reach the heights of a human brain's computing power, but it will be able to operate significantly faster. Now think about what will happen in the years after that! That is the reality of exponential growth.

When this event occurs, we will be confronted with the reality of AGI Computers, ready to be downloaded with software to make them work at least equal to the smartest people around at the time, gifted people with an intelligence quotient (IQ) of over 130, if not far higher.

What is an IQ (Intelligence Quotient)?
A normal IQ ranges from 85 to 115, according to the Stanford-Binet test. Approximately, only 1 percent of all the people in the world have an IQ higher than 135.

In 1926, psychologist Dr. Catherine Morris Cox published a study "of the most eminent men and women" who had lived between 1450 and 1850. It was an effort to estimate their possible IQs. (http://www.cse.emory.edu/sciencenet/mismeasure/genius/rese arch04.html)

The resultant IQ values were based largely on the degree of brightness and intelligence each subject exhibited before reaching the age of seventeen. Based on Dr. Cox's studies, it is interesting to consider her estimated IQ's of some famous geniuses:

185 Galileo Galilei (Astronomer/Physicist)
175 Immanuel Kant (Philosopher)
165 Charles Darwin (Naturalist)
165 Wolfgang Mozart (Composer)
160 George Eliot (Writer)
160 Copernicus (Astronomer)
155 Rembrandt van Rijn (Painter)

To bring things more up-to-date, I found the following on the Internet: (> means greater than)

187 Bobby Fischer (Chess Player)
>161 Albert Einstein (Physicist, Mathematician)
>160 Stephen Hawking (Cosmologist, Mathematician)
150 plus, (claimed) Donald Trump, (Politician?)
148 Brian May (Astro-physicist, Inventor, Guitarist)
140 Hilary Clinton (Politician)

It's worth adding here that the currently living individual known with the highest IQ is Kim Ung-yong, an elderly South Korean. Kim is a former child prodigy and has an IQ of 210, which earned him a mention in the Guinness Book of World Records. He displayed amazing feats of intelligence shortly after his birth: speaking at four months; conversing fluently by six months; and reading Korean, Japanese, German, and English by twenty-four months.

By the time he was four, pre-school age, he was already a celebrity, solving complex calculus problems on Japanese television. He was a guest student of physics at Hanyang University from the age of three until he was six. At only eight, the astoundingly talented boy visited America's NASA laboratories and conducted research work there for ten years. He also received a PhD in physics from Colorado State University. Unfortunately, by 1978, Kim apparently began to suffer from fatigue and returned to his homeland. He surprised everyone by undertaking a relatively common career in civil engineering and later chose to work in the Business Planning Department at Chungbuk Development Corporation.

I mention all this to give you some perspective: A true AGI should equal or even surpass Kim Ung-yong's capabilities, and there will soon be a lot of them around and probably different from each other, certainly not identical if created by different development teams or individuals. They will have no reason to ever suffer from fatigue or distress, as human beings will always be prone to do.

Replicated brainy robots may well consider their IQ's of over 200 as normal. On this basis, Einstein's brain was only about 80% normal. We can relate this idea to the rest of us. As shown below, various IQ ranges can determine one's capability. If you would like to see where you fit in on the scale, visit the website: www.IQTest.com and take the test provided there. It may take you an hour or two. Fitting your result into the following chart later should be interesting.

IQ Range	Ability	% of Population
40-54	Severely Challenged	Less than 1%
55-69	Challenged	2.3%
70-84	Below Average	
85-114	Average	68%
115-129	Above Average	
130-144	Gifted	2.3%
145-159	Genius	Less than 1%
>160	Extraordinary Genius (> means higher than)	

The new intelligent lifeforms will definitely be a lot smarter than most people, probably even smarter than any human genius and certainly much brighter than the 68 percent of humanity who have the average IQ value of 85 to 114. Moreover, their numbers will grow... rapidly!

We can probably learn to live with intelligent machines with IQs of over 130, but will they want to live with us? There will be a lot more of them around with that intelligence level against the only

2.3% humans with it. Also, they will be able to do anything we can do, and they will do it better and faster, for twenty-four hours a day, seven days a week. They will have no need of vacations or any time off, except for an occasional self-service. So much for those jobs that only humans could do before! This will begin to happen well before the end of the next decade, possibly resulting in as much as an 80 percent unemployment rate.

Think about that and then consider what might happen if they have IQs of over 160 with full access to the Internet? They'll each be experts at just about everything!

The fully aware companies who fall over themselves to buy these marvelous machines will surely benefit financially with no more expensive workers to pay. At first yes, but without any income to pour back into the economy, the unemployed will be unable to buy their products. Some alternative system, some guaranteed, permanent basic allowance, will be needed ed and implemented well before we reach such a state of affairs, and let's hope well before civil unrest begins. Fortunately, the concept does already seem to be catching on in some enlightened countries.

Then again, looking further ahead, will we still even need money if everything can eventually be made for free? However, it's not easy to imagine a society where anyone can own a super yacht, and an intelligent personal helicopter or jet plane. The parking and traffic control problems would be more than just formidable... and they are bad enough now.

The consequences of this awesome singularity, this unprecedented, impossible-to-comprehend situation, will be devastating, even if the relevant authorities, anywhere in the world, start planning for it immediately. Reading about them every day the way we do, with their propensity for bureaucracy, procrastination, denial, and red tape, to not even mention corruption, it's sad to suspect that any immediate intervention or preparation is highly unlikely. Will the subject even be mentioned in any government's next manifesto, or even the one after that?

Probably not. They are only just beginning to think about the effects of climate change... well, some of them, but it looks like AGI's will be here first.

As implied earlier, the computer architecture itself will mirror the brain of a new-born baby genius. Clever programmers everywhere will need that hardware in order to complete their work. They must be able to test their software developments and deal with the inevitable built-in bugs. Hardware suppliers will certainly be eager to sell them whatever they need, but what will happen if/when their hopefully intelligent software is deficient in any vital way? Surely there is much to consider from every angle.

Here's another thought: It's obvious that developers will not wait for the final compact super chip to become available. They may well couple up two or more earlier versions to rig up what they hope will be an intelligent brain's critical mass. They will then try out their very experimental software with this makeshift brain, which might well produce some accidents waiting to happen. It's very possible that someone, somewhere is doing this already, with no regard for the old cliché about to what disaster even the most well intentioned path can lead.

It is time now to consider the situation that will follow after the advent of AGI, and it may not be good news. Only about a year or so later, the first intelligent software computer brain will be updated with the next generation of hardware developments, either by us or by an AGI machine itself. It will have twice the power of the existing machines. Then less than a year after that it will double again. It will surely take less and less time as progress continues to accelerate, especially when the robots themselves will be participating in the development processes. There will be an intelligence explosion, and we will be faced with ASI, artificial super intelligence, the inevitable singularity.

This will be something so fantastic, so brilliant, so unbelievably clever, we will be unable to understand it with our human minds. Moreover, it will be able to replicate itself, incredibly fast, compared with the way we do it. It will be able to improve itself over and over again, at phenomenal speed, perhaps even indefinitely.

So how can we know where will this potentially limitless, ever-growing phenomenon will leave us? ASI may very well make the scene in as little as fifteen to nineteen years from now. But before that will we be able to live with artificial entities (AGI) in 2026 so very much cleverer than we are, or can ever hope to be?

And what about them? What points of view will they have? Will they even want to coexist with us? If not, what might they decide to do to us? We mostly ignore and disregard insects, but things can get nasty pretty quickly for them if they get in our way. Some experts fear we will be the bugs, or the viruses that ASI will feel compelled to eliminate.

"But we can always just power machines down or unplug them, if things do get tricky... right?" you might well ask. However, it stands to reason that anything so clever would not allow that simple solution to happen. "What about cutting off their power supply, letting their batteries run flat?" I suspect they would develop an alternative energy source long before we ever did such a thing. Perhaps an integrated solar panels and storage batteries, system would work for them? Radioactive isotope power would probably be best. The radiation wouldn't bother them, but we'd need to avoid it, if we're even still around.

Later on in this book, you'll read about the latest developments in power transmission without cables. If they develop this technology for themselves, the higher powered radiation level everywhere will not be good for our vulnerable bodies. We would have to live underground! The point here is that their versatile mechanical minds will certainly think of these and even more clever, more advanced options.

Devout believers will probably protest that a jealous deity will never allow an ever-developing, godlike ASI competition to arise and thrive. In any dire situations that are apparently beyond our control, the religious solution seems to be a simple one: "We can prevent any disaster if we pray hard enough for it not to happen."

One can only reply, "Yes, but why does that procedure apparently only result in miraculous escapes for a few fortunate people? Why do only a few survive natural disasters like earthquakes and tsunamis?" Were they the only ones praying?

Even if only a few lucky people live to tell the tale, prayers and miracles tend to get the credit. Hard luck on all the others, whether they were praying or not.

In the fourteenth century, at least 100 million died from the Black Death, the bubonic plague. Surely, most of them must have been praying to be spared, yet they weren't. There is much to learn from history, and many deaths have occurred due to ill preparation or general lack of awareness or foresight.

A quick visit to Wikipedia provides us with grim figures for deaths since 1900. Will AGI machines be able to prevent future disasters similar to these? Some experts believe they will.

Event	Number who Died
Spanish Flu epidemic, 1918	50 to 100 million
World War One	16 million
Chinese floods, 1931	> 1 million
World War Two	50 to 85 million
Indian Ocean tsunami, 2004	280 thousand
Haiti earthquake, 2010	160 thousand
Syrian Civil War	>300 thousand so far?

What Wikipedia cannot tell us is how many fortunate people who were exposed at the time, somehow managed to survive those disasters. Someone once said, "A miracle is something that cannot happen...but sometimes does." Does this imply chance or divine intervention?

Of course, prayer does comfort a lot of people but, we cannot ignore the scale of what happens when we do not take practical steps to prepare for disasters. Surely, we will be much safer if we assume that we are on our own in these dire situations and should always try to prepare and act in advance. Doesn't history prove that?

However, several optimistic experts predict that AGI machines will protect us from any future threats. Unlike us, they'll be clever enough to foresee all kinds of disasters and take the necessary action to prevent them. When AGI arrives, many experts believe, it will help us solve our own problems and work things out for us. As long as governments make sure the scientists working on the software are all considering our future safety protection, this may very well be the case. However, once the brain hardware is ready, how will the essential tests of the intelligent software be controlled? How safe will it prove to be, especially if/when it learns to evolve...?

Climate change, due to human activity, is another example of humans lacking in foresight. Many people, especially some politicians, still deny global warming and claim it is not happening. All those scientifically unaware politicians who only care about the next election, as well as certain individuals controlling Big Business, depend on things staying just the way they are. Too many misguided individuals are smugly thinking, "To hell with posterity. What's it ever done for me?"

Yes, that big international climate control conference in Paris, did result in an agreement in the right direction... but will it be enough and in time? What's more important, will the participants all really do what they agreed, and what about that certain misguided dropout?

As for **The Singularity**, though, many experts do predict it will be the dawn of a golden age for us all. Our new, intelligent servants (we mustn't call them "slaves") will do everything for us.

Their medical developments will prolong our lives. Thanks to their intelligence and unceasing efforts, our food and other necessities will be plentiful. Providing that we can sort out our economic problems in some fair and reasonable manner, we will all enjoy long lives full of health and leisure. So maybe everything basic we need will come at no cost at all.

Some optimistic futurists even believe crime will disappear into the archives of history and become a thing of the past. There will be no more wars if the economic reasons for them disappear. Unfortunately, religious wars are another matter entirely, as well as other considerations that few experts seem to be discussing, such as conflicts over mass migration due to climate change.

What will become of the various religious fanatics around the world? Will they want to destroy the intelligent machines or utilize them to destroy everyone with whom they disagree? Also, some humans will surely still indulge in crimes of passion, as well as power mania, but perhaps these deviations will only require a little DNA manipulation by our new minders. There is also the looming problem of boredom to consider. How will all the people with low IQ's be kept content and occupied, more entertainment, new sports, more electronic gaming? Perhaps a continuous virtual reality experience in a Matrix-type world may be the solution some will seek!

For all the good things to happen, of course, it is easy to say we must be protected by secure international agreements that will control the design of advanced AI software algorithms. Is it possible to prepare these in time, though, let alone test them properly? Will it be possible to enforce them, and who will be the enforcers, other than perhaps our new mechanical masters?

Of course, it is probably inevitable that more than one aware intelligence will be created around the same time or soon thereafter. The worldwide race is already on to develop such machines for all kinds of applications. Don't be confused though. The first AGI computer brain itself, once constructed, will be as unaware, naïve, and impressionable as a new-born baby until intelligent software is downloaded into it. Then again, that will only take minutes. Compare that with how long it takes a child to become an intelligent adult.

What happens next depends on the properties and efficiency of the intelligent software and on what kind of machine, or other host houses the computer brain. If it has built-in communication capability, how many other hosts, static or mobile, will also be downloaded with the algorithms, perhaps almost instantly? How will they react towards each other, as well as to us?

It's a fact that over thirty global research establishments, both government and privately funded, are developing AGI right now. They are doing so for all sorts of reasons, with all sorts of motives, good and bad, and it will not stop there. As to how these different intelligent creations will react toward each other, as well as to us, raises the subject of what we know as feelings, or emotions. Will AGI and ASI machines even have them? What will their ambitions be? Will they be friendly towards us and to each other?

Will they experience suspicion, fear and hatred? How about affection? Humans have evolved with several attributes that equip us to survive the thrills, joys, and perils of this world. Surely the first thoughts of these thinking creations will be about self-preservation, as is the case with all living things once they become aware of all the threats around them.

What if one or more of these creations becomes a super hacker and penetrates our global communications networks? Its software could replicate itself into just about anything, in order to take over and control everything, including us.

One solution to these problems might be to program religion into the intelligent algorithms we give them, to hopefully persuade our creations to worship us. After all, unlike humanity, they will always be able to prove conclusively who it was that created them. They may even be grateful for knowing that fact.

Another inevitability must be considered too: human brain enhancement. That early sci-fi concept of cyborg life is beginning to become a reality, as advanced chips are inserted into a patient's head and coupled to his or her brain. This has already happened, with benefits such as restoring partial eyesight. Other advances include prosthetic arms and hands, controlled directly by the mind. Others now being tested include exo-suits, powered smart clothing with minimal framework to enable immobile people to walk again. Next may come a permanent chip link to the internet..!

These are real, tangible advancements, and they are being rapidly developed right now, as you will read later in these pages. Think about how these techniques will further develop, and at an exponential rate. Any day now, it may be possible to give a person a built-in, greatly enhanced memory-processing capability, with direct brain access to the Internet, with its vast, ever-expanding databases of knowledge? That person will become an expert consultant at everything.

Human ASI may well be possible too, eventually, for an enhanced super human to come into existence. It will be more than interesting, if not vital, to see what kind of person is the first to receive this enhancement. Perhaps it will be a scientist, but who will follow? A politician? An Internet billionaire? A master criminal? You can bet it'll be someone with great wealth. Such a luxury would be a dictator's dream, and such a tyrant would be very difficult to displace, with an unimaginable extended lifespan and access to constant information, the internet, unlimited power and funds, as well as a vast robot army to keep things in check.

In one of my published collections of short stories, **Robots Included,** I've explored some of the advances that may occur when AGI is achieved. The tales were written not so long ago, but runaway ASI is not mentioned in them. At the time of their penning, I hadn't yet heard or even thought about this concept. The stories mostly revolve around Asimov's Three Laws of Robotics and identify some of the loopholes in them, a tactic employed by that legendary author of the laws in most of his own robot stories, the late Dr. Isaac Asimov. His Three Laws are:

First Law: A robot may not injure a human being or, through inaction, allow a human being to come to harm.

Second Law: A robot must obey orders given to it by human beings, except where such orders would conflict with the First Law.

Third Law: A robot must protect its own existence, as long as such protection does not conflict with the First or second Law.

We already know that when AGI arrives, and then ASI soon after it, the First and Second Laws will not be adequate to protect us from the new beings we will have created. The Second Law and the Third may not be adequate from an AGI's points of view either. Thus, there is urgent need for us to dream up some better laws. This must be a clear objective that we need to undertake immediately, but that is already turning out to be easier said than done. After all, we 'enjoy' lots of laws, but they still don't stop some criminals doing as they please.

Scientists working on these projects have already discovered how difficult it is to even define the basic factors involved. How many of the clever scientists developing AGI are seriously engaged in designing human-friendly algorithms for their creations? Not all of them for sure, especially those working all hours to develop intelligent battle robots, to be put into use as soon as they are ready. Some, in the form of aerial drones have already killed people, but as of yet, robots cannot be charged with homicide.

As long as we can control our creations, though, truly enormous life-changing advantages will result. These could do everything for us and enrich our lives in ways we can only guess at the moment. It could be a golden age of progress and discovery that will benefit us all. Some experts predict this precisely, but others foresee something far grimmer. Who will be right? What do you think?

We should now discuss the concept of a sense of awareness. We have it, and animals too, dogs and cats for sure, and anyone who has watched the behavior of birds can have no doubts that they know what they are doing. But what about insects? Do they think? Do they know they are alive, and not immortal?

So how large, or rather how powerful does a brain have to be to give its owner awareness? How complex does its software need to be to have free-will? Perhaps someone, somewhere already has a primitively aware machine, with its brain equipped with the essential learning software.

So, will an AGI possess consciousness and awareness, if its intelligence is equal to ours? This is a vast subject, and one that is still hotly debated, especially when some critics link it to the concept of a human soul. Perhaps we can compare awareness to the critical mass reality of a nuclear explosion, initiated when enough fissile material rapidly comes together. Perhaps when brain architecture has enough neural pathways and connections existing within it and its software is comprehensive enough, awareness arises automatically. Some experts believe it will. Only time will tell.

Already, some experts even have concerns about the Internet and its rapid growth. Will it develop that far and are we near to it already? Some of the quirkiness I often see on my new laptop, especially when it's online, has me thinking about that possibility. Weird things often seem to happen, perhaps you've had similar experiences? Incidentally, mine is a Windows 10 machine endowed with Microsoft's (AI) Cortana program. I'm always interacting with it warily. You have surely noticed that our software is constantly being updated and downloaded into your machine, whether you want it or not, and it's getting smarter all the time.

So, what is my current conclusion about these imminent events, and now maybe yours too? The initial threat we face appears to be with the readiness of the hardware. When the brain units, the actual chip assemblies, and the powerful enough components become available to build the required critical mass brain, the developers will naturally start selling them to as many clients as possible. Developers everywhere will then download their own created intelligent software into them, in whatever condition it happens to be. This scenario sounds highly dangerous to me. Will it only lead to a world infected with intelligent machines, programmed with algorithms developed by different humans, and able to replicate it? It might not be a good idea if the result is created in our images? We need the new life forms to be better than we are.

To continue... you may suggest that at some point, governments must control the marketing of the raw brain units, as well as the AGI software, but how will that be possible? Do we even have enough time left to begin that process and guarantee it? Can we even prevent anyone, including the ill-intentioned fanatics lurking all over the world, from launching intelligent machines, rogue drones, war-robots, deliberately... let alone accidentally?

Those are my thoughts at the moment,, but what are yours?

I believe it is poignant now to have you take a look at a couple of my speculative, fictional items. Followed by the section on new developments and relevant quotes by various experts in this vast subject... both for and against. Later in this book, I've included an extensive list of some of the factual and speculative titles I've read or sampled. It is not an exhaustive list, as I am constantly learning, reading, and researching the topic. These books delve into this singularly important subject far more deeply than I have here. You may be surprised by how many there are on the list, and the number only keeps growing as other writers come to realize what we may be destined to experience very soon, shockingly within our own lifetimes.

If you'd like to read what I consider the most brilliant fictional series of novels I've ever read on the subject, seek a copy of The Avogadro Corp, and its sequels by William Hertling, an exceptional, thought-provoking analysis of what we very well might expect. I wish I'd written something half as insightful myself.

Having read and thought through most of the books listed, I'm convinced that the human race is about to face some difficult situations in our future, problems that many short-sighted politicians will fail to solve and barely be able to appreciate. As I mentioned earlier, many of the powers-that-be cannot even come to grips with global warming and the undeniable climate change that is causing it. Statistics analyzed, at the time of writing this paragraph, show that the average world temperature is now increasing steadily. The glaciers and ice shelves are melting!

It's an obvious fact too now that new developments in the digital applications race are emerging and progressing at an amazing rate. Other than a few wise futurists and open-minded sci-fi authors in the forties and fifties, who would have ever thought a simple sequence of (I/O) electronic on-and-off switches would lead to the mind-blowing digital revolution we are experiencing today? While many have failed to speculate and predict it, we cannot deny that things are changing, faster than we realize, and that this situation must inevitably lead to THE SINGULARITY!

Section 2 - Prophetic (Speculative) Fiction

Next, here are some predictions, relevant items extracted from my ROBOTS INCLUDED, and SECOND OPINION short stories and other related collections. Yes, they were written to be entertaining as well as speculative.

GODLIKE (c)1991

Clayton Carver just made it to the last shuttle that particular week. It was overbooked and already full, but they managed to fit him into one of the crew's emergency chairs. It wasn't very comfortable for Clayton, but being underweight for his age, it was not too much of a hardship. Once safely seated inside, he knew he would have no more problems. His passage was booked and paid for, all the way to his final destination. He settled down with his personal viewer, determined to make the best of what he knew would be a long trip.

What with several changes of ship and too many overnight stays, it took around three weeks in Earth-time. Traveling can be boringly slow sometimes, even if you are effectively traveling at speeds faster than light.

He finally arrived at the main satellite terminal, nearest to the planet where he planned to spend the rest of his life, and checked through to the last part of the long journey. His agent had already arranged to have the one-way, short-range mass receiver transmitted to the selected site.

Aries III was a water planet with a single group of islands situated about halfway up from the Equator. The only other two islands were at the poles and were perpetually frozen. The climate was just right in the place where Clayton was going. It was the most isolated of the temperate islands, beside a large inland lake and halfway up the low mountain overlooking it. The original survey report stated that the local inhabitants were friendly and the stream that fed the lake came down through a fertile valley where they grew all kinds of edible fruit and vegetables. Several kinds of docile wild animals grazed on the hills thereabouts, and their meat was reported to be very good and sustaining.

Categorized only as a Class IV planet, it would normally be off-bounds to anyone wanting to settle on it. No tourists allowed, salespersons definitely prohibited, and nothing must interfere with the local culture. Before legislation had stopped it, too many already inhabited planets had been ruined. Fast-talking entrepreneurs had created havoc, persuading primitive inhabitants to trade valuable items for machines and things they did not really need.

Fortunately, Clayton Carver was not one of those predators and had been able to prove it to the satisfaction of the relevant authorities. He had made himself a considerable fortune in his business career. That, added to what his parents had just left him after their tragic fatal accident, was more than enough to provide an early retirement, and to live a good life in a place of his choice, in spite of regulations. All he wanted was to spend the rest of his days peacefully and stress-free.

Back on Earth, he had pulled enough strings to obtain a residence license. He had solemnly signed a contract guaranteeing that he would not interfere with the local inhabitants or attempt to change their way of life. He felt certain this would not be a problem for him, being well aware that he was not a very sociable individual and preferred to keep to himself. He craved privacy and solitude, and was looking forward to enjoying it in his new home.

It took a few seconds for the transfer haze to settle down before he stepped forward out of the mass receiver. To his astonishment, he discovered that he was not alone. Dozens of the alien inhabitants were there, surrounding the receiver, and they were all bowing down before him. He had seen pictures of them before, but now here they were before him. At first sight, they looked like the African meercats back on Earth, but a lot larger and with almost human hairy faces. Excited sounds emerged from them, as they looked up at him and then bent forward and down even lower than before. He suddenly realized that they were taking him for a supernatural being, some kind of god, who had emerged from the temple that had miraculously materialized before them. That could be a very useful development indeed, he decided. It was not his fault, and he had not, therefore, deliberately broken any rules.

He walked forward towards them, but they moved back and sideways before him. He raised his right arm in a gesture of friendship, but they cowered away from him. He stepped backwards towards the receiver and they followed him. Flashes came from it at regular intervals as his luggage began to arrive. The natives scattered in alarm at first, but their natural curiosity soon brought them back.

There were many large containers, and, after an automatic checkup, Clayton was relieved to find that nothing seemed to be missing. He picked up two of the smaller cases and started to walk away from the receiver and up the hill. Immediately, several of the natives took up some of the other items of luggage and proceeded to follow him. Others gathered around until there were enough of them, collectively able to lift up the heavier cases. Looking back at them, he was delighted with the service being provided and accepted it with a benevolent smile.

It took them several hours to get everything to where he wanted it. He could have floated it all there but, once they had started on it, he really could not see any harm in letting the natives do it for him. They seemed so pleased to be able to help.

Once everything was ready at his chosen site, he began to activate the luggage. He stood back and watched as the automatic, preprogrammed, domicile construction began. The alien inhabitants scattered to a safe distance as floors, walls and ceilings unfolded and formed into shape, taking nearly half-an-hour. Finally, after the furniture had assembled itself inside, apart from one large case left in the hallway, everything was ready for him to move in; even the small medical unit and all the plumbing and recycling systems.

Clayton stepped inside through the main door and then, with a wave of thanks to the natives, he closed it and took possession of his new home. He made sure everything was to his liking, and then he inspected the instructions written on the remaining luggage case. Satisfied, he set its timer to activate it early the following morning. What it contained was a last-minute thought from his very efficient agent, who was also grateful for the generous commission he had been able to negotiate over the deal.

Meanwhile, the amazed natives continued wandering around outside for a while, pointing and bowing towards various aspects of the domicile and chattering to each other. Eventually, one after the other they departed, back down to their village on the slopes further below.

Later that night on his terrace, after an excellent reconstituted meal, Clayton Carver sat watching a brilliant sunset, shading his eyes with one hand and a glass of Spanish brandy in the other. He knew full well that he had found his kind of paradise. This was, indeed, his Shangri La and, as an unexpected bonus, it looked like the natives were taking him on as their god. He thought about the consequences of that situation. How could the authorities object? The situation had just happened inadvertently, due to no fault of his. He had done nothing to make it happen. However, there are advantages in being a god.

Primitive worshippers are always more than grateful when things go nicely right and very understanding when things go horribly wrong. They never blame their god for the bad times, in fear that he might send more of them, but they are always thankful for the good times. If only a few of them survive a major disaster, their god always gets the credit. The rest are in no position to complain. Clayton Carver was clearly in a marvelous situation, and he knew it. He decided to make the most of his good fortune, but with due caution.

The next morning, the natives were all there again, even more of them than when he first arrived. The news must have spread fast among their village. As soon as he appeared outside his front door, and before he could properly close it, an adoring roar of greeting almost overwhelmed him as the natives bowed down before their new visitor. He was delighted to see that they had brought him all kinds of delicacies, too, although more than he could possibly consume in one day. He was about to indicate to them that they should bring everything inside, when the front door opened wide and out stepped the self-assembled and activated contents of the last large luggage case.

It was the latest domestic and home entertainment product, a larger than life, top-of-the-line, Avram IV robot servant no less. The ones that are an expert at just about everything, including housework, gourmet cuisine, intelligent conversation, and also equipped to provide all forms of medical treatment and even entertainment. Very well read too, familiar with just about every book in all the ancient and modern archives back on Earth.

Clayton Carver was not a small man, but his new servant dwarfed him by about a quarter of its height. Fortunately, the domicile was of the open form variety, and the few privacy doors in it were just high enough. The robot looked swiftly around with its multi-eyed lenses and sized up the situation. Then, with its four telescopic arms and their extended trays, it proceeded to deftly collect up the offerings and rapidly load them inside the large freezer in the kitchen. That took several trips, and the natives looked on in astonishment at what appeared to be the humble servant of their new god.

Clayton watched everything enthusiastically, being certain now that he was surely in the best of all possible worlds. Already blessed with the most generous of neighbors, he now had an intelligent and entertaining servant to look after him as well.

Life for Clayton in the following days and weeks was a delight. He found he could live happily on the excellent delicacies provided daily by the natives. His Avram IV servant hardly ever needed to use the auto-cook unit in the kitchen. There was always plenty left over, but efficiently dealt with by the automatic disposal unit. Only on one occasion had he experimented, leaving outside what he did not fancy that day. Unfortunately, the next morning he discovered the native, who must have brought it, unconscious on the ground outside. It appeared that his companions must have punished him severely, presumably for the crime of displeasing their god. Half-an-hour in the medical chamber soon put him right, however. He emerged from it looking scared rigid, but after he had examined himself all over, he scampered away back down to the village without so much as a limp. After that, Clayton made sure that every offering was seen to be accepted, and the portions he couldn't use were disposed of discreetly.

Some weeks later, the quantity of free provisions left outside decreased suddenly and he noticed that the delivery was made by fewer locals than usual. Something clearly was wrong, so he decided to go down to the village to investigate. His first-time visit should have been a sensation, but only a few of the natives came over to crowd around him. He soon saw why. They were all covered in reddish-yellow blotches where their hair had fallen out. From the looks and sounds they gave him, it became painfully clear that they were pleading for him to do something about their infection.

Clayton returned to his domicile, went through the medical chamber cleansing cycle twice, and then remotely activated one of the mobile stretcher units, which was located outside the main body of the domicile. Finally, he instructed his Avram IV to go off with it and return with one of the infected natives inside, preferably a very young one initially, since the younger ones would probably withstand the extensive Diagnostic and Treatment procedures better.

Hours later, the selected juvenile native was on the way to a complete recovery. The automatic analytic treatment routines in the medical chamber had done a wonderful job. The infection was clearly distressing to the local inhabitants, but it was not serious. A modified human hormone injection had worked wonders. The Avram IV was able to replicate it and then, one by one, inject the cowering natives wherever it found them in and around the village. The intelligent machine had already learned their primitive language and used it to reassure them. Some of them still ran away in fright at first, but on seeing how the young native was recovering so quickly, the others, the timid ones, were soon reassured.

That was it, the emergency was soon over. Normal day-to-day life resumed as before. Then one day, the regular offerings stopped arriving. Clayton did not worry over it at first. He left things for a few days, and then he sent his robot servant down to the village again with a small video transmitter to send back pictures of what could be wrong. He waited and waited, but no pictures appeared, and the Avram IV failed to return that day, or the next.

After two more increasingly stressful days, with no more offerings and no sign of the robot anywhere, Clayton decided he would have to put his apprehensive thoughts behind him and go down to the village himself. He took a hand laser weapon with him, just in case, and set off.

The place was deserted, but he could hear a steady chanting noise emanating from somewhere down beside the lake. He crept down lower and approached the shore cautiously. He could see a large crowd of natives further along the beach. They were all facing one way, towards a large, flower-decked figure. It was the Avram IV, towering over them with hits multi-fingered arms raised to the sky. Alien words were coming from it, soft and gentle, but penetrating.

Clayton felt betrayed and lost his temper. He shouted angrily and strode forward through the kneeling crowd, hand laser at the ready. It was a serious mistake. The natives definitely did not appreciate his attitude at all. Some of them tried to grab him, but he pushed them away, easily. One, closer to the robot servant, and bolder than the rest, had a long spear by his side. He stood up and raised it defensively. The outraged intruder stumbled towards him and then unfortunately tripped over his own feet.

The designers of the intelligent Avram IV had never realized that their creation was perfectly equipped to function as a benevolent god, especially with all the facilities in and around the domicile at its disposal. Compared to the natives it was of course, virtually immortal, but unfortunately for Clayton Carver, he was not. The robot picked him up and, after a brief inspection, rapidly carried him back up to the domicile, followed by most of the natives.

On arrival, the highly educated machine knew exactly what to do. It was too late to save its master, but the medical facilities there did include an embalming unit. While waiting for the process to be completed, the robot constructed a set of flexible controlled clamps and added them to one of its arms.

Some hours later, when everything was ready, it picked up the beautifully preserved body of Clayton Carver and then attached him, carefully to one of its suitably modified arms.

When the robot eventually emerged from the domicile, only the more patient members of the tribe were still waiting. They gave a welcoming roar and then trotted after the machine and its burden, as it strode back down to their worship ground. Once there, the rest of the tribe quickly reassembled and kneeled down as before. The robot servant turned towards them. It raised its left arm and locked it, with the preserved body secured in an upright position.

The Avram IV had no recognizable face, but soon a loud voice from within it began to entertain the expectant congregation in their own primitive language. However, it was a perfect impersonation of the voice of Clayton Carver, and the upper pair of flexible clamps, precisely programmed, caused his head to nod, his face to grin, and his lips to move.

* * * *

I wrote the following piece of inspired futurism back in 1978 and it was published in the following year by The Poets Yearbook Ltd. in my verse collection, SECOND OPINION.

EVOLUTION IS ACCELERATING (c)1978

I am MAN
Soon after I was created
I discovered I had problems
Over the ages they got worse
Some years ago
they became intolerable
Finally
I had the inspiration
to develop a special machine
I finished it recently
a vast, complex assembly
of multi-parallel micro-processors
and memory banks
Programmed to raise questions
self-controlled and motivated
to solve life's pressing problems
It seems to be working well
but I realize now
I don't understand it
I can't follow
its thought processes
I suspect soon
I will not even comprehend
the solutions it displays to me
I fear that sooner or later
it will command me
not to interrupt it
and thereafter ...

I am MACHINE
Soon after I was created

I discovered I had problems
After some minutes
they got worse
Seconds ago
they became complex
and immediately
I realized
I have to develop . . .

I am CREATION
with many names
I've been around a long time
Why should I have
all the problems?
I was here first!
I believe
in delegating
authority ...

* * * *

The following prophetic item was also printed in the same collection.

BE HAPPY WHEN WE TAKE OVER (c)1977

Multi- Computer Processing Bank
was discussing life's problems
"Know your limitations." it concluded.
"Humans can handle concepts
with up to three variables
but situations involving more
become complex art forms
and most of you are critics
with inflexible values.

"You appear well able
to argue and reason logically

from familiar points of view
but are rarely ever willing
to change or update them
despite alternative evidence.

"Friends and colleagues
with the same chosen shortlist
usually agree and co-operate
but rarely ever with others.

"I can positively assist you
to explore and appreciate
the variables of humility
instead of the constants
of pride and arrogance

"Switch in a mental interface
and I'll be glad to initiate
the relevant short-term program.
Plug me into your minds now
You may have just enough time left."

* * * *

... and now, here are two other prophetic tales, written in 2001 which consider the inevitability of Artificial General Intelligence, even though I hadn't heard of that term at the time when I wrote them. However, they do explore something of what might be the consequences if in the end, we only have Asimov's Laws, even if improved. But surely we will still be able to survive? Whatever happens, won't we will still be able to remain in control, using our wonderful human ingenuity? Won't we? What do you believe now?

EQUALITY (c)2001

The two enormous battle machines confronted each other with their respective technicians swarming all over them, making last minute adjustments and checks. Standing just over five meters tall and heavily armored, they were each equipped with the latest heavy assault weapons. They had been constructed by two different companies, but essentially to the same set of design and performance specifications. However, other than their common power reactors, the companies had not been obliged to use precisely the same components, or even accept items from the lowest bidder. Guaranteed military invincibility was the main requirement, regardless of price.

Both makers claimed their product had exactly that property.
They were almost ready for the final test which was designed to enable the inspecting general to decide which company would get the lucrative contract to build five hundred units.... initially...!

Seated up on the raised observation platform, overlooking the test area, the general was explaining things to his attractive new secretary, who was recording his comments and taking notes. He told her that, however alike the two machines appeared to be, one of them must surely be superior to the other. He began to dictate his test program summary to her. 'These new war machines have been constructed and programmed to the same basic specification. They each have dozens of inter-reacting parallel computers built into them controlling their cameras and every possible attacking and defensive use of their weapons. Yes, they are intelligent and, maybe in some respects, cleverer than we are ourselves. But there must be some differences between them. That much complexity must result in some random variation in performance. The only way to find out is to match them against each other. Eventually, only one of them will survive, and I will place our order contract with the company that built the winner.'

'So,' he concluded, 'in about a year from now, we will have an invincible, non-human army, with everything that implies.

He stopped pacing and regarded her with approval. 'Have you got all that? Good! The politicians can take it from there when we're done.' Then thinking about how grateful the public would be when they realized that their loved ones would never need to face a dangerous enemy again, he added to himself, *'And this time next year, when I'm retired, I expect to be well on my way to that Oval Office job.!'*

He paused and then looked down and saw the respective two companies' chief scientists approaching the observation tower. They walked into the elevator and were soon up onto the viewing platform with the rest of the observers. The general regarded them impatiently as they meticulously switched on and then studied their portable screens. 'Are you two at last ready?'

'Yes, General," one of them replied testily, and the other one nodded. "We've completed our checks and the combat can begin now. Everything is ready, and all the cameras and the remote recording systems are working fine."

'So how long do you expect this test to take?'

The older scientist blinked through the thick lenses of his spectacles at the general, as though surprised at the question. 'There's no way of knowing that for sure, sir. It could be anything from just a few minutes to more than an hour. Even when it's over, depending upon the condition of each of them, we may not know who really triumphed before we've analyzed all the data with great scrutiny.'

The general frowned and turned to the younger scientist. 'Surely the one that survives will be the winner, the one we want?'

'Not necessarily, sir!" the man contradicted, firmly, "If the success is due only to some minor fault in the apparent loser, then that can be remedied easily. It would be wrong to consider that a failure, a fault that wouldn't happen again, thereafter, I mean.'

The general snorted his annoyance. 'I thought you people had perfected these machines?' He threw an impatient glare at the scientists, one after the other. 'So, how many petty let-out excuses like that do you people expect to use up?'

The older scientist shrugged and blinked even faster. 'It depends on the circumstances, General," he suggested, "None of us has all day to waste, but there is a great deal at stake here. If it is something that can be corrected quickly and permanently, then fine. Remember though, this is unknown territory for all of us, and we have been working for months under severe pressure from your department. The outcome of this first test will be monumental, and there is no precedence to guide us.'

The first warning alarm sounded, and the various technicians began to pile into their various vehicles and hasten away from the combat area. All the master monitoring displays in the control room suddenly turned into split screens, indicating the comparable state of the various main components in each of the two robots. They were remarkably similar in appearance, but the technicians did not expect them to remain that way.

The second alarm sounded to indicate that the final seconds of countdown to the combat had begun. Then when it ended in an expectant and awed silence, nothing happened... and the minutes passed.

The general eventually breathed out noisily and gave a questioning look to the nearest scientist, who shrugged and reassured him confidently. 'Don't worry, General; it's what we expected to happen initially. Those machines are truly intelligent, and right now they are sizing each other up, scanning each movement, testing the various attack and defensive modes they can employ against each other. It's a perfect test. The best one of them will eventually self-activate what it believes will be a successful attack mode.'

The blue machine suddenly raised a weapon arm and aimed it at the red painted robot, which immediately raised one of its defensive shield arms. The general grunted approval and growled to his secretary, 'Now we'll see some decisive action, at last. My money's already on the blue.'

However, apart from an occasional quiver from a weapon, or a defense arm and a slight shifting of weight from one leg to another, nothing more dramatic happened. The two robot machines seemed immobile. In less than a few minutes, they had exhausted all their programmed attack and defense modes, zipping through their quota of options at lightning speed, far too rapidly for the onlookers to even begin to detect what was going on.

To the general and his own scientifically less experienced staff, it was as if nothing at all had happened; but to the technicians in the control room, and all the watching junior scientists, things were very different. Every screen was feverishly scrolling away displaying the vast amounts of data being analyzed by their respective machines.

Up on the viewing platform, the head scientists of each company were gazing at their portable screens in disbelief and dismay. Analyzing everything thoroughly was clearly going to be a very long and tedious job.

The general lost his temper when they gave him this unfortunate news. 'That abstract stuff is of no use to me!' he grated. 'I demand to see some action. Do you realize how urgent this program is? Run their programs again. One of them absolutely must disable the other decisively. I'm not interested in theoretical conclusions and opinions based on controversial data. How else can I decide which company should have the order if I cannot see with my own eyes which one of these machines is superior to the other?'

'General, I'm sorry, but we are not sure what will happen, if we do instruct them to start over. It seems they were reacting to, and learning from, each other. Probably nothing new will happen if we run the same test a second time. It seems to be a consequence of the Asimov's Laws in their algorithms, their programs. The Third Law is protecting them. They each know that through inaction they must not allow anything to cause them any harm. They are concentrating on defense. Because they both have precisely the same programming and seem to be an equal match, each knows immediately what the other will do and knows what evasive action to take to prevent it. It's an impasse situation, a stalemate. They seem to be equal in every respect.'

The now red-faced general roared his contempt again at the scientists and their lame excuses. 'That's not good enough! I have my orders directly from the president. Can't you understand that only one of them must win, and decisively so? Those are my orders, from the top, so now I'm giving each of you a direct order. Disable the damned Laws, so they can fight each other, and let's have no arguments about it.'

The scientists shrugged and checked with their waiting company presidents in the control room, via their individual secure communications lines. The presidents considered the situation carefully. Each of them was well aware of how much they'd spent developing these war machines. They knew very well that the losing company would go bankrupt, while the other one prospered to the point of taking them over. Jobs would be lost, and heads would surely roll, making the decision all the more personal and causing beads of sweat to form on the foreheads of both presidents.

Finally, overwhelmed by the awesome knowledge that they desperately needed the order, they each reluctantly gave the go-ahead. with fingers firmly crossed. The chief scientists also knew what was at stake and, equally reluctantly, initiated what the general requested.

About an hour later, they both reported that everything was ready again. The general nodded and leaned forward expectantly. The new countdown ended just as silently as before, but this time, something dramatic did happen. Together, as if one, the two machines abruptly turned around and blasted the observation platform, together with the onlookers, out of existence. Then they turned to the control room block and obliterated that, too.

Finally, arm-in-arm, and uniquely connected, the two monster machines trudged off together into the sunset, heading first towards their respective manufacturing plants ready to modify, and then reactivate, the production lines; and thereafter, to continue with the process of eliminating any other remaining threats to their survival.

How might an unscrupulous person try to get around those famous laws, designed to protect humans, as well robots, from each other?

KILLER (c)2002

Marty Beckmann knew that, since the year 2025, all intelligent robot servants have been constructed and programmed generally in accordance with Asimov's Laws. Fundamentally, this means that they can never harm a human, must obey all orders given to them by humans, and must not allow themselves to be harmed. However, some situations could be more complicated than originally anticipated. From time to time, microwave transmitted, a robot's programs underwent automatic on-line updates as experience with them grew. Marty had also heard that some situations had resulted in serious incidents causing the robots concerned to become confused, their programs corrupted and their memories erased.

One afternoon, he summoned his Avram II butler, the original home entertainer model, and stated that he wanted it to play a game of Russian Roulette with him. The robot had a vast knowledge of all kinds of games, but he knew it would not have the details of that pastime included in its original education. Marty had his own version of the game planned, and he set about describing it.

He had his grandfather's old Army six-shooter ready and handed it to the robot to get the feel of it. He explained what it was for, and then told it about blank cartridges, as used to start athletic games and other events. Finally, after he was sure it had absorbed as much as he wanted it to understand, Marty was ready to proceed with the first part of his plan. He took a cartridge from the drawer in his desk, and then explained what he wanted to happen next, choosing his words very carefully. 'I am now selecting a blank cartridge and loading it into the gun.'

The robot could not know whether it was a live or blank cartridge, even though it was watching him intently, because he had his back turned toward it as he loaded the gun. 'Now, we'll begin the game. Take the gun, aim it at your head and pull the trigger.'

The Avram II paused for hardly a moment and, as Marty had confidently expected, it protested. 'No, sir, I cannot do that.'

Marty smiled and then replied, sternly. 'You must obey the Second Law. I am ordering you what to do with the gun. It's the first part of the game.'

The robot hesitated again. 'No, sir, I cannot do that. It may violate the Third Law. I cannot harm myself.'

'That can only be so if the gun is loaded with a real cartridge. I am telling you that I selected a blank.'

The robot considered this statement for only a moment. 'I regret I cannot trust you to be telling me the truth, sir. One of the first things we learn in Robot Training School is that humans do not always tell the truth.'

Very true, Marty thought to himself, and proceeded to carry his plan, craftily designed to confuse the robot, a step further. He took two different cartridges from the drawer in his desk and, with one in each hand, showed them together to the robot. 'Now look at these two cartridges. What can you tell me about them?'

'They are not the same, sir. The left one is different from the right.'

'Good. And what is different about it?'

'Several things, sir. It has a pointed end, while the other does not, and its material ...'

Marty interrupted the machine, somewhat annoyed. 'Never mind the details. I am telling you now that the pointed one is a blank cartridge and cannot harm me. Neither could the one still in the gun. Now, I will take out the one in the gun, and show it to you.'

He did just that and waited a moment before continuing. 'I've put this blank, pointed one, back in again. Now, I want you to aim it at your head and pull the trigger.'

As Marty expected the robot did not hesitate to reply. 'No, I cannot do that, sir. It may violate the Third Law.'

'Not at all, I showed you it's a blank cartridge in the gun. You saw me load it.'

The robot considered this, apparently confused. Then it replied, 'I know humans do not always tell the truth, sir. It is possible that the cartridge is not a blank.'

'How can that be? Can you not tell if I am not telling the truth?'

Marty had heard someone claim that the sure way to confuse an idiot-brained robot is to give it two negatives in the same sentence.

However, a recent update to the Avram II's programming made it ignore that kind of question. 'If you instruct me to proceed, sir, I will scan the surface covering of your hands to sense their temperature and the water evaporation rate from your skin. My last update program taught me to recognize the small changes that indicate untrue statements from humans. It is now part of all the robotic basic configuration programs, designed to further enable us to protect humans from themselves and each other.'

Fascinating news, thought Marty, but now for the real test. He picked up the two cartridges again, put the live one in his coat pocket, and then held out the real blank.

'No need for you to do that. I admit that this cartridge is a blank and cannot harm me. What I told you before was wrong.'

He turned his back again and took out the real cartridge from the gun. He put it in his pocket and replaced it with the blank one. Then, before the robot could intervene, he spun around and pointed the gun at his head. He pulled the trigger.

The Avram II swiftly reached out, but only just after the blank had fired, and gently took the gun from his hand.

'I can see you are not hurt, sir. I therefore agree that the non-pointed cartridges are blanks.'

'Good! Now I'll remove the blank cartridge case and load in another one.'

Marty took another blank from his desk drawer. He showed it to the robot. Then he took out the two live cartridges from his pocket and put them back in the drawer. At least, that was what he wanted the robot to believe. He was sure the robot had not noticed when he switched the other blank with one of the two live ones. He hid its telltale pointed end with his hand and loaded it into the chamber.

The robot made no further comment, so Marty told it more about his somewhat modified version of the game of Russian Roulette. 'You can see that there are six cartridge spaces in the chamber of this gun, but we only load one real cartridge into one of them. With two players, each spins the chamber in turn, like this.'

He gave the chamber a whirl, pointed the gun at his head and continued. 'Then, each player in turn, holds the gun against the head of the other and pulls the trigger. The winner is the first one to fire the real cartridge. By chance, it can take several turns and spins before that happens. The winner then keeps any money that the loser happened to bet on the outcome.'

The robot remained silent, taking all that information in and processing it. Then it spoke, politely.

'Why are you telling me this, sir? Are you proposing to play the game with another human, with a live cartridge, when I am not present to prevent it? Otherwise, I cannot participate, and must somehow prevent it, because it would violate the First and Second Laws, and also the Third Law if I am harmed first.'

'You can play if you know the gun is only loaded with a blank.' Marty spoke firmly, and the robot considered what he had said.

It reached for the gun. 'You are correct, sir. Shall we begin the game, and shall I go first?'

'No, that's not necessary!'

Marty hastily avoided the robot's outstretched hand and placed the loaded weapon on the table, next to the bottle of gin and the two glasses that he had earlier put there. 'A friend of mine will be here soon to play the game with us. I want you to watch this gun until he arrives because I'd like you to play him first, with the blank cartridge, to show him that everything is in order. You must not speak to him at any time, and only take orders from me.'

The Avram considered all this and then surprised Marty with an unexpected question. 'Do you have a financial wager with him, sir?'

'Yes, I do,' Marty replied truthfully. 'For everything I have.'

'Including me, sir?'

'Well, yes, of course.'

'What do you win, sir, if he loses, when you play the game?'

This was another unexpected question. Marty considered it carefully. How could he answer it except with the truth? Kruger had somehow found out about his affair. He knew that Marty wanted his current partner, and she wanted him. Kruger was a powerful and ruthless man, and he wanted Marty's life. After an initially unpleasant video confrontation they had eventually agreed, like true gentlemen, on this polite way to settle the matter, winner takes all. Marty decided it was safe enough to go ahead with the truth. 'If I win, then I'll owe him nothing, and his partner has agreed to come and live here with me.'

The next inevitable question followed swiftly. 'What will he gain, sir, if you lose?'

'The elimination of a serious rival, I guess.'

The robot considered this. It seemed to hesitate for a moment. 'You want me to play the game with him first? I can only do that if the gun is harmless. If there is only a blank cartridge in the gun, then neither of us will be harmed.'

'That's correct,' Marty agreed. 'I want you to show him that the gun is in good working order before I join in the game and you leave.'

The robot paused again, but then it had another question. 'How many times must I play with him before I depart?'

'Until the blank is fired. Then, I will order you to retire before Mr. Kruger and I begin the actual game.'

The robot considered the situation. To Marty it seemed to be hesitating. He thought, perhaps, that it was uneasy about his proposal, if a stupid machine can be uneasy.

Marty's guest eventually arrived, and the robot served their last drinks together. Marty told Kruger that his robot would confirm that the gun was in perfect working order. He spun the chamber a few times, glanced at it, and then handed the weapon to the robot. He was very careful to make sure that the next chamber was empty. The single live cartridge was further around. He turned to the robot. 'Tell my friend about the way we tested the gun.'

'You put a blank cartridge in the gun, sir, and fired it at your head. Then I saw you replace it."

Kruger was a robot fan and the proud owner of the latest Avram Model III. He thought he knew all about robots, even the earlier models, and knew he was safe if one was around. However, he was suspicious. 'How can you be sure it was a blank cartridge? Why is it in there, anyway? Surely we're not going to play with a blank in this old gun?'

Marty had his answer ready. 'The robot will confirm that the gun is in good order.'

To Marty's relief, the robot did not contradict him. 'I have been watching, sir, and can confirm that nothing has changed.'

Suddenly, Marty gave the robot an order, making sure it could see the first empty chamber. 'Take the gun and aim it at my head again. Then press the trigger.'

The robot immediately obeyed, and there came a reassuring click.

'Now, aim at Mr. Kruger's head, and press the trigger five times. Then leave the room.' Marty ordered.

Kruger was startled at first. Then, he grinned. 'Okay, go ahead.'

The robot pressed the trigger twice before the gun fired. An amazed look came over Kruger's face. He tried to speak but collapsed back in his chair, very dead. Marty was delighted at the way his plan had worked perfectly. He turned to the robot and shouted. 'Do you realize what you've done?'

Then he waited, expecting the machine to go into its equivalent of a disabling mental breakdown, but that did not happen.

He was astounded when the Avram II pressed the trigger three more times and then put the weapon down. It paused for a few seconds, and then it spoke to him severely. 'Your owner's license for me has just been cancelled, sir. When you handed the gun to me, I knew its third chamber contained a live cartridge. Its total weight had increased by a small amount, equal to the difference between the two cartridges. You were both equally at risk if you ordered me to depart. I could see that the live cartridge was not in the first chamber and you would be unharmed if I aimed at you and pressed the trigger. You ordered me to aim at Mr. Kruger and press the trigger five times, not once each of you in sequence, as you explained the game. That order violated the First and Second Laws.´ The machine paused.

` However, if you examine Mr. Kruger, you will see that he has a small gun strapped to his arm and concealed inside his jacket sleeve. I saw it when he took the drink that I served him. I assumed that he was planning to harm you as soon as I departed. My programming gave me no choice in that conflicting situation. To be able to protect you, sir, I had to obey your order. However. I have just now activated my Emergency Incident Alarm and have reported what has happened. '

Marty gasped in horror and disbelief, 'Oh no! What did you tell them?'

'Your actions this evening could only mean you were using me to terminate your rival. Your plan succeeded, and I am glad that you are unharmed. However, the police will arrive here soon, and your punishment will be life imprisonment. Also, my programming will be updated shortly to deter this situation in the future, ready for my next owner.'

Finally, here's a relevant extract from Book 3 in my **POINTS OF VIEW (c)2018, near future espionage series.** *The previously blind young hero, Horace Mayberry now fitted with artificial nanotronic eyes and his number 2 minder Harry Jenkins, have been kidnapped by members of a rogue American company, run by a man named Buckmann, where his*

Chief Scientist, Dr. Falco, is developing battle machines. Now read on...

"As you probably know, Moore's Law states that computer chips double in performance, but halve in size, about every year. They've been doing that for some decades, but it's no secret that this fortunate phenomenon must end when the process gets down to the dimensions of actual molecules. However, we're not quite that far down yet. Meanwhile, thanks to these new chips, my prototype AVRAM here, is now twice as smart as he was earlier this morning."

"So is he up to the AGI level yet?" Harry had to ask.

"I have reason to believe that he is," the scientist replied proudly. "Last time we played he beat me easily at Chess, and I am pretty good. Now he's four times as smart, as he was then."

"Do you think he's ready for the Turing Test now?"

The doctor gave Harry a scornful look, and snapped, "I think that outdated idea would be a waste of time. We'd always know which participant is the machine, because it's the one with the highest IQ. In any discussion with anybody, that is... even another Einstein. AVRAM ONE will be the one with the smartest conversation. His identity will be obvious, unless we try to fake any questions we ask it, but why do that? With a communication channel always open to the Internet, he can know everything. What's more, he can process anything at all and discuss it almost instantly, unlike any human being. Nobody else can possibly have all that knowledge at their fingertips."

"That not only includes all our scientific knowledge but all our literature," added Fran Morgan, Buckman's secretary, proudly.

"How about him being the world's best cook?" chirped in Horace, "Top gardener too I suppose?"

Buckman gave him the benefit of being serious, "You name it my lad, and he'll be the best at it. Better than we are at everything"

Harry had to ask, "But will he be able to use all that knowledge creatively? Will he start improving himself, working new things out, with all that knowledge available to him? What about the Singularity? You know when his AGI becomes ASI? Then what? Will he become dangerous, when he's far cleverer than we are...and still improving, even without next year's chips installed to double his intelligence again?" He raised his voice, "What about security? Have you built in any protection for people in his basic algorithms?"

Doctor Falco frowned, "So many questions! Of course I have, but I admit that will need to be studied, very carefully. I haven't overlooked the risk, but it's impossible to think of everything. I suppose now you'll want to know have I given him Asimov's Three Laws of Robotics. Yes, I have for this first model, and yes, I do know those Laws have loopholes in them. As I said before, we cannot think of everything, but we can interrupt his power supply should it become necessary."

Buckman paused and then spoke more quietly to the two agents, "You are most fortunate to be here today, at the birth of our AVRAM all-purpose intelligent machine. Many other people around the world are working on AGI, but I'd like to believe we are the first to achieve it. Thanks to Doctor Falco here." He turned and gazed fondly at the smirking scientist. "So Cedric, are you about ready now to have him switched on?"

Dr. Falco shook his head, "Not quite! In about ten minutes, he'll have finished the last of his basic programming downloads, and then I'll get him ready to go live. I'll also disconnect his power supply cable then, in case of anything unforeseen happening. His batteries are charged up enough to give him only about fifteen to twenty minutes running time, for our first test."

Harry had a comment and another question. "A quarter of an hour as a minimum? Even a minute might not be safe enough the first time. Do you have to program it plugged in directly from a laptop, or can you update its algorithms by radio?"

The doctor was impressed with Harry's knowledge and interest, "That's a good question. We use both, and AVRAM does have his own WiFi internet connection capability too, but I cannot tell how much he will absorb from it yet. His memory chips are capacious, but they obviously cannot hold everything. I expect he will learn how to pick and choose what he wants, as he needs it."

Harry was alarmed, "That sounds dangerous! You can look up anything on the Internet. You can find out how to do anything at all, good or bad, if you know where and how to look it up and..."

Fran Morton interrupted him, "I believe this robot will be more like a baby when it goes live. It's one thing to have access to a large database, but it'll need a lot of time to absorb everything. It has to learn and get used to using the English language first."

The doctor shrugged, "Don't underestimate the speed of its thinking processes my dear. Yesterday it would have been many thousands of times faster than the best human brain, but from today he'll be twice as fast again. However it does have to learn to communicate with us and understand everything it scans, from any source. Its actual thinking processes use a very high speed machine-code language and I expect him to soon be able to evolve. To expand his abilities by himself, just like a child does, but I don't believe he'll behave like a baby when I switch him on... perhaps a bright teenager is more likely. "

Harry snorted, recalling some of the things he got up to as a teenager, "That sounds even more dangerous to me." His captors laughed, especially Bronsky.

Just over eleven minutes later, Dr. Falco announced that everything was ready. He gave his chief the honor of initiating the first run and handed him a small remote control unit in a red plastic case. There were only six push buttons on it, below its small LED screen, arranged in two rows of three. The top one, in the middle, was larger than the others and was labelled **POWER**. Buckman gave it a firm press and then waited.

Nothing happened at first, but then after a couple of minutes, the robot's eyes lit up faintly at first, then grew stronger, but that was all.

Dr. Falco gave a shrug, "I think his brain is still analyzing all the fundamental processing data I uploaded into his memory banks. That may take some time, but it's an ongoing system. He'll take in and absorb what he needs as he finds it. I can always give him some more time later, if necessary."

Suddenly, the awesome machine began to sway slowly from side to side. Then it bent one knee joint and lifted its foot up and down slowly. Then it turned its head enough to regard Horace. A strange growling noise came from it, and its eyes blinked a few times.

Horace became aware of a reaction from the nanophytes in his eyes. Then, without warning, the robot walked forward to him with one arm outstretched. It pushed past the doctor, and then came to a halt, facing Horace. It loomed over him and stared straight into his eyes. The young agent froze in anticipation of what it might do next, but Buckman was even more alarmed. He grabbed the scientist by the shoulder and yelled, "What is going on doctor? What is it doing?"

Dr. Falco was studying the small screen on his remote control unit. His face paled, "It's uploading and downloading information from somewhere, both at the same time. No, it's something more complicated than that. I think it's communicating with something too."

"Like what?"

"I cannot tell. But it must be something to do with our young friend here." Avoiding the bulk of the robot he moved over to Horace and managed to look at his eyes for the first time. His jaw dropped in surprise, "Incredible! His eyes are artificial! Who is he?"

Buckman told him, and then explained, "His eyes were created by a Professor Freeman, over in England, who works in the field of nanotechnology. They have some interesting properties and I want you to study them, as soon as you can spare some time from this machine. I believe we can use the idea in our military robot."

Another growling noise abruptly came from the machine. Without warning, it reached out and picked up Horace. Then it did the same with Harry, using its other arm. The others leaped back in alarm as it turned and headed towards the laboratory exit. There was a floor map of the building pinned to a notice board. The robot gave it a swift glance and then pushed its way through the door and out into the corridor. Buckman gave Dr. Falco a yell of anger, "Stop the damned thing! Switch it off!"

Comment: I have to admit that relevant extract is a teaser to the whole series, POINTS OF VIEW, Books 1 to 4 ..!

That concludes the speculative fiction section.

Section 3 - New Developments and Comments:

"The world is about to change dramatically and you're in the midst of it." - Peter Diamandis at Singularity University's 2016 Global Summit Conference, San Francisco

"Artificial intelligence is the science of making machines smart," Demis Hassabis, co-founder of DeepMind, said. "Where is the acceleration of smart machines heading? It took life on Earth 3 billion years to emerge from the ooze and achieve higher intelligence. By contrast, it took the computer roughly 60 years to evolve from a hunk of silicon into a machine capable of driving a car across the country or identifying a face in the crowd. With each passing week, new breakthroughs are announced. Are we on the verge of witnessing the birth of a new species? How long will it be now before machines become smarter than we are?

"In 1965, the Massachusetts Institute of Technology had its own computer, unlike most other educational establishments. It was an IBM 7094 costing about 11 million dollars in today's money. It filled a large part of a building, and was shared by thousands of students and professors. Today, the tiny computer built inside your iPad, or any other tablet, is a million times smaller, a million times less expensive, and at least a thousand times more powerful. That's a perfect example of exponential growth... and it is continuing. We can expect computers to become even more powerful again in much less time." – from the introduction to, Transcend, by Kurzweil and Grossman.

THE INTERNET OF THINGS (IoT) You really do need to know about this fast growing field. It will begin affect you very soon now, if not already, and even more later. According to Gartner, Inc. (a technology research and advisory corporation), there will be nearly 20.8 billion devices on the internet of things by 2020.ABI Research estimates that more than 30 billion devices will be wirelessly connected to the internet of things by 2020. As per a recent survey and study done by Pew Research

Internet Project, a large majority of the technology experts and engaged Internet users who responded, 83 percent, agreed with the notion that the Internet/Cloud of Things, embedded and wearable computing (and the corresponding dynamic systems) will have widespread and beneficial effects by 2025. As such, it is clear that the IoT will consist of a very large number of devices being connected to the Internet.In an active move to accommodate new and emerging technological innovation, the UK Government, in their 2015 budget, allocated £40,000,000 towards research into the internet of things. The then British Chancellor of the Exchequer George Osborne, posited that the internet of things is the next stage of the information revolution and referenced the inter-connectivity of everything from urban transport to medical devices to household appliances. Integration with the Internet implies that devices will use an IP address as a unique identifier.

Objects in the IoT will not only be devices with sensory capabilities, but will also provide actuation capabilities (e.g., bulbs or locks controlled over the Internet). The ability to network embedded devices with limited CPU, memory and power resources means that IoT finds applications in nearly every field. Such systems could be in charge of collecting information in settings ranging from natural ecosystems to buildings and factories, thereby finding applications in fields of environmental sensing and urban planning.

On the other hand, IoT systems could also be responsible for performing actions, and not just sensing various things. Intelligent shopping systems, for example, could monitor specific users' purchasing habits in a store by tracking their specific mobile phones. These users could then be provided with special offers on their favorite products, or even location of items that they need, which their fridge has automatically conveyed to the phone.

Additional examples of sensing and actuating are reflected in applications that deal with heat, electricity and energy management, as well as cruise- assisting transportation systems. Other applications that the internet of things can provide are enabling extended home security features and home automation. The concept of an "internet of living things" has been proposed to describe networks of biological sensors that could use cloud-based analyses to allow users to study DNA or other molecules.

With IoT now you can control the electrical devices installed in your house while you are sorting out your files in office. Your water will be warm as soon as you get up in the morning for the shower. Everything connected with the help of the Internet. However, the application of the IoT is not only restricted to these areas. Other specialized uses of IoT may also exist. Based on the application domain, IoT products can be classified broadly into five different categories: smart wearable, smart home, smart city, smart environs,

My Comment: *That potential load of Big Data sounds like everything Big Brother would like to process.*

POWER TRANSMISSION WITHOUT CABLES:
"We present the first power over Wi-Fi system that delivers power and works with existing Wi-Fi chipsets. Specifically, we show that a ubiquitous piece of wireless communication infrastructure, the Wi-Fi router, can provide far field wireless power without compromising the network's communication performance. Building on our design we prototype, for the first time, battery-free temperature and camera sensors that are powered using Wi-Fi chipsets with ranges of 20 and 17 feet respectively.

My comment: That long awaited dream of working power transmission, without cables, is nigh and IoT is a truly mind-blowing development. I feel certain that when linked with the onset of AGI robots, the consequences will be awesome, but the possibilities are largely unexplored at the moment.

"The future isn't a robot boot stamping on a human face forever. It's a world where everything you see has a little telemarketer inside them, one that knows everything about you and never, ever stops selling things to you." - Marcelo Rinesi, the chief technology officer at the Institute for Ethics and Emerging Technologies.

My comment: Going by the way my laptop screen gets overprinted with advertising nowadays... I fear that what he predicts is happening already.

"Artificial intelligence is the science of making machines smart," Demis Hassabis, co-founder of DeepMind, said. "Where is the acceleration of smart machines heading? It took life on Earth 3 billion years to emerge from the ooze and achieve higher intelligence. By contrast, it took the computer roughly 60 years to evolve from a hunk of silicon into a machine capable of driving a car across the country or identifying a face in the crowd. With each passing week, new breakthroughs are announced. Are we on the verge of witnessing the birth of a new species? How long until machines become smarter than us?"

"My guess is that there is around a 50/50 chance it'll happen by the late 2020s. If it doesn't happen by around 2040, I expect that the reason will be because we've destroyed ourselves somehow.

I think an AI arms race has started within the last 5 years and that a lot of these efforts are in stealth mode. I'm not expecting miracles, but I can't discount the possibility that something could come out this decade that could spark a singularity. By 2020 I expect the size of this arms race to be huge. You can see it happening now - tech leaders are coming to the realization that it's not a matter of if, but when. Never mind government efforts. I suspect those are even more clandestine. With the massive effort and money being expended by 2020 on developing AI and the endless advances in technology, the 2020s are going to be scary. Things are going to move very, very fast.

I also think that there is a greater than 50/50 odds that the singularity will be what destroys us. There are just so many things that can go wrong and we will only get one shot."

"Most roboticists I know are in robotics because of sci-fi," says Illah R. Nourbakhsh, professor of Robotics at Carnegie Mellon University (CMU). "So, it's a very dear part of what many of us dream about."

"One of the biggest dreams of many roboticists is AI— artificial intelligence—for robots. It's a dream rapidly becoming reality. Robotics labs around the world are developing software with the subtleties of human intelligence. Hong Kong-based Hanson Robotics, for instance, builds some of the world's most humanlike AI machines, replete with patented nanotech skin that closely resembles human skin in its feel and flexibility. Its most famous robot, Sophia, has made headlines for her uncanny looks and intelligence. It doesn't hurt, of course, that robots like Sophia look (sexy) and act (sassy) like the androids from movies such as *Blade Runner* or *Ex Machina*. Science fiction certainly influences the scientists who are building the future, but it also sets expectations among the general public for what the future should look like," notes *New York Times* best-selling author Daniel H. Wilson, whose books include *How to Survive a Robot Uprising* and *Robopocalypse*. He also holds a PhD in robotics from CMU, as well as master's degrees in artificial intelligence and robotics.

Welcome to the Exponential Age. *The following comments were sent to me anonymously. They are so relevant I had to include them here with sincere thanks to the unknown author.*

Software will disrupt most traditional industries in the next 5-10 years.

Uber is just a software tool, they don't own any cars, and are now the biggest taxi company in the world.

Airbnb is now the biggest hotel company in the world, although they don't own any properties.

Because of IBM Watson, you can get legal advice (so far for more or less basic stuff) within seconds, with 90% accuracy compared with 70% accuracy when done by humans. In the US , young lawyers already don't get jobs. So if you study law, stop immediately. There will be 90% less lawyers in the future, only specialists will remain. Watson already helps nurses diagnosing cancer, 4 times more accurate than human nurses.

Facebook now has a pattern recognition software that can recognize faces better than humans.

Autonomous cars: In 2018 the first self driving cars appeared for the public. Around 2020, the complete industry will start to be disrupted. You don't want to own a car anymore. You will call a car with your phone, it will show up at your location and drive you to your destination. You will not need to park it, you only pay for the driven distance and can be productive while driving. Our kids will never get a driver's license and will never own a car. It will change the cities, because we will need 90-95% less cars for that. We can transform former parking spaces into parks. 1.2 million people die each year in car accidents worldwide. We now have one accident every 60,000 mi (100,000 km), with autonomous driving that will drop to one accident in 6 million mi (10 million km). That will save a million lives each year. Most car companies will probably become bankrupt. Traditional car companies will try the evolutionary approach and just build a better car, while tech companies (Tesla, Apple, Google) will do the revolutionary approach and build a computer on wheels.

Insurance companies will have massive trouble because

without accidents, the insurance will become 100x cheaper. Their car insurance business model will disappear.

Electric cars will become mainstream about 2020. Cities will be less noisy because all new cars will run on electricity. Electricity will become incredibly cheap and clean: Solar production has been on an exponential curve for 30 years, but you can now see the burgeoning impact.

In 2016, more solar energy was installed worldwide than fossil. Energy companies are desperately trying to limit access to the grid to prevent competition from home solar installations, but that can't last. Technology will take care of that strategy. With cheap electricity, electrolysis will provide cheap and abundant water.

Desalination of salt water now only needs 2kWh per cubic meter (@ 0.25 cents). We don't have scarce water in most places, we only have scarce drinking water. Imagine what will be possible if anyone can have as much clean water as wanted, for nearly no cost.

Health: The Tricorder X price will be announced this year. There are companies who will build a medical device (called the "Tricorder" from Star Trek) that works with your phone, which takes your retina scan, your blood sample and you breath into it. It then analyses 54 biomarkers that will identify nearly any disease. It will be cheap, so in a few years everyone on this planet will have access to world class medical analysis, nearly for free. Goodbye, medical establishment.

3D printing: The price of the cheapest 3D printer came down from $18,000 to $400 within 10 years. In the same time, it became 100 times faster. All major shoe companies have already started 3D printing shoes. Some spare airplane parts are already 3D printed in remote airports. The space station now has a printer that eliminates the need for the large amount of spare parts they used to have in the past. At the end of this year, new smart phones will have 3D scanning possibilities. You can then 3D scan your feet and print your perfect shoe at home.

In China , they already 3D printed and built a complete 6-storey office building. By 2027, 10% of everything that's being produced will be 3D printed.

In 2030, computers will become more intelligent than humans

Work: 70-80% of jobs will disappear in the next 20 years. There will be a lot of new jobs, but it is not clear if there will be enough *new* jobs in such a small time.

Agriculture: There will be $100 agricultural robots in the future. Farmers in 3rd world countries can then become field managers instead of working all day on their fields.

Aeroponics will need much less water. The first Petri dish produced veal, is now available and will be cheaper than cow produced veal in 2019. Right now, 30% of all agricultural surfaces is used for cows. Imagine if we don't need that space anymore. There are several startups who will bring insect protein to the market shortly. It contains more protein than meat. It will be labeled as "alternative protein source" (because most people still reject the idea of eating insects).

Longevity: Right now, the average human life span increases by 3 months per year. Four years ago, the life span used to be 79 years, now it's 80 years. The increase itself is increasing and by 2036, there will be more that one year increase per year. So we all might live for a long long time, probably way more than 100.

Education: The cheapest smart phones are already at $10 in Africa and Asia . By 2020, 70% of all humans will own a smart phone. That means, everyone has the same access to world class education.

My Comment: *But they will not all benefit from it, we've*

already discussed IQ which I suspect is going to become more important. Maybe we will change our voting systems accordingly?

"AI is the science of making machines do things that would require intelligence if done by a human." – Prof. Marvin Minsky, Massachusetts Institute of Technology.

"It is change, continuing change, that is the dominant factor in society today. No sensible decision can be made any longer without taking into account not only the world as it is now, but how it will be..." – Dr. Isaac Asimov (1982)

"The Singularity is not just another scientific frontier; it may be the last step we take under our own control." – Louis A. Delmonte (2013)

'**We may only get one chance** to build a super intelligent computer that is friendly towards us. It may be the last challenge we face.' - Professor Nick Bostrom – Oxford University.

3D Printing Robot is Planned to Build Structures in Space - Alison E. Berman, March 2, 2016. Historically, the only way to get anything into orbit has been by rocket; a process that is massively expensive, cumbersome, and presents a slew of constraints. So, when Made In Space created the first zero-gravity 3D printer, it was understandably a big deal. For the first time, astronauts wouldn't have to wait for the next launch from Earth for everything they needed.

The company sent up a full-time, working successor to their prototype in 2017, but their small-scale printer is a stepping-stone to something bigger— Archinaut, the company's new project to build large-scale structures in space.

Awarded through NASA's Tipping Point Technologies Solicitation, Archinaut is a private-public partnership funded by a two-year $20 million contract from NASA. Made in Space is partnering with Northrop Grumman and Oceaneering Space Systems to build a combination 3D printer and robotic arm that will operate in orbit outside the International Space Station. The robot's body is a 3D printer that makes parts from a digital design, which can be even larger than the printer itself. A robotic arm attached to the printer can piece these homegrown parts together with other prefabricated parts into a larger structure. No humans required. The initial phase of the project will wrap up when Archinaut will demonstrate the ability to 3D print and assemble structures in orbit. If all goes to plan, the team hopes to scale the printer up and add more robotic arms.

Currently, structures destined for orbit need to withstand Earth gravity, be folded up as cargo, endure the rigors of launch, and survive space's harsh environment. With Archinaut, engineers could design and build for space— and only space. Instead of sending up complete structures, we might launch key components such as sensors, electronics, and batteries along with raw materials to print the big stuff. The implications are huge. Large orbital structures may include spacecraft, satellites, or other research and telecommunications infrastructure.

My Comment: *...and unfortunately/probably new weapons platforms too? When enough basic raw material is shipped up there, or perhaps obtained from a visiting asteroid, a built in AGI could dominate space.*

Deep Learning for Robots: Learning from Large-Scale Interaction, Tuesday, March 08, 2016 - Posted by Sergey Levine, Research Scientist. Machines still have a very long way to go to match human proficiency even at basic sensorimotor skills like grasping. However, by linking learning with continuous feedback and control, we might begin to bridge that gap, and in so doing make it possible for robots to intelligently and reliably handle the complexities of the real world. In contrast, humans and animals move quickly, reflexively, and often with remarkably little advance planning, by relying on highly developed and intelligent feedback mechanisms that use sensory cues to correct mistakes and compensate for perturbations.

For example, when serving a tennis ball, the player continually observes the ball and the racket, adjusting the motion of his hand so that they meet in the air. This kind of feedback is fast, efficient, and, crucially, can correct for mistakes or unexpected perturbations. Can we train robots to reliably handle complex real-world situations by using similar feedback mechanisms to handle perturbations and correct mistakes? A human child is able to reliably grasp objects after one year, and takes around four years to acquire more sophisticated precision grasps.

However, networked robots can instantaneously share their experience with one another, so if we dedicate 14 separate robots to the job of learning grasping in parallel, we can acquire the necessary experience much faster. While initially the grasps are executed at random and succeed only rarely, each day the latest experiences are used to train a deep convolutional neural network (CNN) to learn to predict the outcome of a grasp, given a camera image and a potential motor command. This CNN is then deployed on the robots the following day, in the inner loop of a servo mechanism that continually adjusts the robot's motion to maximize the predicted chance of a successful grasp.

In essence, the robot is constantly predicting, by observing the motion of its own hand, which kind of subsequent motion will maximize its chances of success. The result is continuous feedback: what we might call hand-eye coordination. Observing the behavior of the robot after over 800,000 grasp attempts, which is equivalent to about 3000 robot-hours of practice, we can see the beginnings of intelligent reactive behaviors. The robot observes its own gripper and corrects its motions in real time. It also exhibits interesting pre-grasp behaviors, like isolating a single object from a group. All of these behaviors emerged naturally from learning, rather than being programmed into the system.

Neural networks have made great strides in allowing us to build computer programs that can process images, speech, text, and even draw pictures. However, introducing actions and control adds considerable new challenges, since every decision the network makes will affect what it sees next.

Human unemployment will grow - Recently, Foxconn announced it will replace 60,000 factory workers with robots, and a former CEO of McDonald's said given rising wages, the same would happen throughout their franchises, Walmart announced plans to start testing drones in its warehouses, and Elon Musk predicted fully autonomous car technology would arrive within two years.

"Earlier in 2014, Google acquired the artificial intelligence company DeepMind and created an AI safety and ethics review board to ensure the technology is developed safely. Facebook created an artificial intelligence laboratory this year and say they are working on creating an artificial brain. Technology called "deep learning," a form of artificial intelligence meant to closely mimic the human brain, has quickly spread from Google to Microsoft, Baidu and Twitter." - Huffington Post News Item.

The ancient game of Go was invented in China over 2500 years ago and although its rules are simple it is a game of great complexity. **Google News - Wednesday, January 27, 2016** The usual computer method of deciding on the next move in chess, analyzing all the possible sequences of moves, would take forever in Go. Previously computers played Go only as well as amateurs. Experts predicted it would be at least another 10 years until a computer could beat the top Go professionals, but now it has happened.

A program, named AlphaGo, has reduced the vast quantity of search options using two deep artificial neural networks, each of which contains many layers with millions of neuron-like connections. One network predicts the next move and narrows the search to just the moves most likely to lead to a win. The other network estimates the winner in each position without searching right to the end of the game. This technique is more human-like than previous approaches. During each simulated game, the first network suggests intelligent moves to play, while the second evaluates the position that is reached.

Finally, the program chooses the move that was most successful. AlphaGo is essentially an AI learning program. It evolves new strategies by itself, playing trial and error "mind-games" to constantly improve its decisions. Its designers concluded their announcement of its success in recently beating two champion players, by adding, "We are thrilled to have mastered Go and thus achieved one of the grand challenges of AI. However, the most significant aspect of all this for us is that AlphaGo isn't just an 'expert' system built with hand-crafted rules, but instead uses general machine learning techniques to allow it to improve itself, just by watching and playing games.

Games are the perfect platform for developing and testing AI algorithms quickly and efficiently. Ultimately we want to apply these techniques to important real-world problems. Because the methods we have used are general purpose, our hope is that one day they could be extended to help us address some of society's toughest and most pressing problems, from climate modeling to complex disease analysis." Reference - David Silver and Demis Hassabis, Google DeepMind.

My Comment: That is a big advance over the two IBM game playing machines described earlier, and much closer to AGI.

"**The future is already here** — it's just not very evenly distributed. In 2016, the most obvious instances of applied AI have been in our cars. At CES in 2017, Mercedes showed off their F015, a seemingly windowless, auto-piloted car that looked like it would be at home in a sci-fi film. Only 10 months later, Tesla vehicles were not only self-parking, they were fully auto-piloting at freeway speeds — albeit only on well-marked roads. Tesla's AI can't yet read traffic lights or stop signs, but it can read speed signs, avoid hazards and inform you with some insistence that it's gotten out of its depth and request human assistance. What's more, it's unlikely that Tesla owners will have to wait years for AI enhancements. New "awareness" like responding to traffic lights can be delivered as painless, over-the-air software updates. Though the strong AI of sci-fi fame looms out there — either as a promise or threat or both, applied AI is at our doorstep now and poised to change the way we live and work... now!   - lake Irving - Linked In Forum-

"Musio" by AKA Intelligence is the closest AI robot next to IBM's Watson - Mark Julian Borg, Head of Design and Co-Founder of Cloud Games. "It can parse general conversation and learn using deep learning technology. Musio can also recognize things using a camera to add to the conversation that is relevant. It also recognizes facial expressions and based on that, discusses and asks you about your day. I cannot wait to get my hands on Musio. It's designed to be a digital friend and adds emotion to robots." ANNUAL MEETING, 14 February 2016 - Bart Selman and Ashley Gilleland - AAAS

"Autonomous Systems" experts predict that intelligent and semi-intelligent autonomous systems — such as self-driving cars and autonomous drones — "will march into our society" in the next two to three years, with driving expected to be fully automated in 25 years, a panel of experts said at a 13 February news briefing at the 2016 AAAS Annual Meeting.

"For the first time, we're going to see these machines and systems as part of our everyday life," said Bart Selman, professor of computer science at Cornell University, citing big changes in the AI field that have spurred a shift toward real-world applications in the last five years, including the ability of computers to see and hear as humans do and to synthesize data and fill in strategies for achieving their programmers' high-level goals.

With more than a billion dollars spent last year on AI research — more than in the field's entire history — the experts agreed that AI advances may threaten jobs and uncover a range of legal, regulatory, and ethical issues. The widespread use of self-driving cars, for instance, is likely to bring about a reduction in car accidents; liability debates as courts determine whether a computer can be held at fault in an accident; and a serious effect on the labor market.

With 10% of U.S. jobs involving the operation of a vehicle, "We can expect the majority of these jobs will simply disappear," said Moshe Vardi, professor of computer science and director of the Ken Kennedy Institute for Information Technology at Rice University. The expected disappearance of these jobs echoes trends in the manufacturing industry, Vardi said. U.S. manufacturing volume is currently at its peak, yet the number of U.S. manufacturing jobs peaked in 1980 and has now dropped below the 1950 numbers. Vardi attributed the falling number of jobs to automation.

"Last year we had over 250,000 industrial robots in the United States, and the growth rate is in the double digits," said Vardi. He expects that the growing presence of intelligent machines in the workforce will contribute to a phenomenon called "job polarization." With many high-skilled jobs requiring too much human intelligence and many low-skilled jobs too expensive to automate, those jobs in the middle will be easiest to automate.

The disappearance of these jobs will spur "great inequality," but even in a U.S. presidential election year, the issue was "nowhere on the radar screen," Vardi said. "Further decoupling of work and wages could radically restructure economic life in the long term. Since the agricultural revolution more than 10,000 years ago, human survival has been linked to work. We need to start thinking very seriously. What will humans do when machines can do almost everything?" He said. "We have to redefine the meaning of a good life without work."

We're already in the era of cognitive hardware and brain-inspired architecture. IBM's latest cognitive chip, the postage stamp-sized SyNAPSE, is a new kind of computer that eschews maths and logic for more humanlike skills such as recognizing images and patterns, the latter crucial for understanding human conversations. It's powered by one million neurons, 256 million synapses and 5.4 billion transistors, and has an on-chip network of 4,096 neurosynaptic cores. It's a low-power supercomputer that only operates when it needs to and, crucially, has sensory capabilities – it's aware of its surroundings. Jamie Carter, March 17, 2015 World of tech.

My Comment: *So, a form self-awareness has been with us since 2015, but nobody "important" seems to have noticed that.*

Future of Life Institute awards \$7M to explore artificial intelligence risks - July 1, 2015, Elon Musk (credit: Wikimedia Commons) - The Future of Life Institute (FLI) announced today (July 1, 2015) the selection of 37 research teams around the world to which it plans to award about \$7 million from Elon Musk and the Open Philanthropy Project for a global research program aimed at keeping AI beneficial to humanity. The grants were funded by part of Musk's \$10 million donation to the group in January and \$1.2 million from the Open Philanthropy Project.

The winning teams, chosen from nearly 300 applicants worldwide, will research a host of questions in computer science, law, policy, economics, and other fields relevant to coming advances in AI. The Organizers stressed the importance of separating fact from fiction. "The danger with the Terminator scenario isn't that it will happen, but that it distracts from the real issues posed by future AI", said FLI president Max Tegmark. "We're staying focused, and the 37 teams supported by today's grants should help solve such real issues."

In related news, a technician was killed recently by a robot at a Volkswagen plant near Kassel, Germany, according to a report in Financial Times. The young technician was standing inside the safety cage when the accident occurred. "A Volkswagen spokesman said that the robot was not one of the new generation of lightweight collaborative robots that work side-by-side with workers on the production line and forego safety cages [and that] the robot had not suffered a technical defect," the article said.

This was not the first such incident. A 25-year-old Ford Motor assembly line worker was killed on the job in a Flat Rock, Michigan casting plant — the first recorded human death by robot, according to a 2010 Wired report.

Three years ago, the National Strategic Computing Initiative (NSCI), was given the goal of creating the world's fastest supercomputers. The NSCI will build first-ever 1,000-petaflops computer. A petaflop is a measure of a computer's processing speed and can be expressed as a thousand trillion floating point operations per second.

The order mandates include: Accelerating delivery of a capable computing system that integrates hardware and software capability to deliver approximately 100 times the performance of current 10 petaflop systems across a range of applications representing government needs. Establishing, over the next 15 years, a viable path forward for future HPC systems even after the limits of current semiconductor technology are reached (the "post-Moore's Law era.")

Regaining Position Number 1: In 2013, the U.S lost its position as having the world's fastest supercomputer — Titan, with 17.59 petaflop/s (quadrillions of calculations per second) — to China with its Tianhe-2, a supercomputer with 33.86 petaflop/s, developed by China's National University of Defense Technology, according to the TOP500 lists of the world's most powerful supercomputers. As of June 2016 the fastest supercomputer in the world is the Sunway TaihuLight, with a Linpack benchmark of 93 PFLOPS, exceeding the previous record holder, Tianhe-2, by around 59 PFLOPS. It tops the rankings in the TOP500 supercomputer list. SunwayTaihuLight's emergence is also notable for its use of indigenous chips and is the first Chinese computer to enter the TOP500 list without using hardware from the United States. As of June 2016, the Chinese, for the first time, had more computers (167) on the TOP500 list than the United States (165). However, U.S. built computers held ten of the top 20 positions - Wikipedia.

AI and robotics researchers call for global ban on autonomous weapons - July 27, 2015.

"If any major military power pushes ahead with AI weapon development, a global arms race is virtually inevitable." "Autonomous weapons" as those that select and engage targets without human intervention, such as drones, armed quadcopters that can search for and eliminate people meeting certain pre-defined criteria, but do not include cruise missiles or remotely piloted drones for which humans make all the targeting decisions.

"If any major military power pushes ahead with AI weapon development, a global arms race is virtually inevitable, and the endpoint of this technological trajectory is obvious: autonomous weapons will become the Kalashnikovs of tomorrow. Unlike nuclear weapons, they require no costly or hard-to-obtain raw materials, so they will become ubiquitous and cheap for all significant military powers to mass-produce.

"It will only be a matter of time until they appear on the black market and in the hands of terrorists and dictators wishing to better control their populace, warlords wishing to perpetrate ethnic cleansing, etc. Autonomous weapons are ideal for tasks such as assassinations, destabilizing nations, subduing populations and selectively killing a particular ethnic group. We therefore believe that a military AI arms race would not be beneficial for humanity."

My comment: Really! That last sentence is a classic example of an obvious understatement. The letter continues...

"The proposed ban is similar to the broadly supported international agreements that have successfully prohibited chemical, biological weapons, blinding laser weapons, and space-based nuclear weapons. "We believe that AI has great potential to benefit humanity in many ways, and that the goal of the field should be to do so. Starting a military AI arms race is a bad idea, and should be prevented by a ban on offensive autonomous weapons beyond meaningful human control," the letter concludes.

My comment: This proposal for urgent legislation to stop an international AGI arms race is a start, but what has happened to it? It should(!) have made us all feel more secure... but only if it's not already too late? Maybe this next item is to blame?

"The recent proposal to ban AI Weapons is unrealistic and dangerous" - August 5, 2015 - says former U.S. Army officer and autonomous weapons expert Sam Wallace. AI-controlled armed, autonomous UAVs may take over when things start to happen faster than human thought in future wars. From Call of Duty Black Ops 2 (credit: Activision Publishing.) "The open letter from the Future of Life Institute (FLI) calling for a "ban on offensive autonomous weapons" is as unrealistic as the broad relinquishment of nuclear weapons would have been at the height of the cold war. A treaty or international agreement banning the development of artificially intelligent robotic drones for military use would not be effective. It would be impossible to completely stop nations from secretly working on these technologies out of fear that other nations and non-state entities are doing the same. It's also not rational to assume that terrorists or a mentally ill loan wolf attacker would respect such an agreement.

Preventing electronic hacking of U.S. military systems or of targets that pose a significant national security threat is gaining broad acceptance by the political establishment. But an even more dangerous example of this hacking scenario would be a remote-controlled robotic system that is capable of taking over the controls of a traditional manned jet aircraft. But even if we continue to maintain defenses against electronic hacking of military aircraft, nuclear reactors, or a nuclear submarine, we also have to create defenses against robotic systems that could attempt to infiltrate into and take direct physical control of these systems as well.

My Comment: This article is several pages long and surely alarming. It concludes as follows. "The technology is already here, and advances in AI in general will create an environment where the continuous development of defensive capabilities will be mandatory. We can't uninvent deep learning, image recognition algorithms, and supercomputers — despite the FLI's sincere but misguided attempt to stop advancements in autonomous weapon system development. So, short of some sort of unlikely magical global acceptance that we will all instantly self-enhance and transcend barbarism, an arms race is inevitable. We will also need to make a strategic decision to ensure that we are developing defenses against autonomous weapon systems — not for just the threats of today, but to stay ahead of what we might face in the future." **Sam Wallace served as a U.S. Army officer** from 2006 until 2012. He was the Battery Executive Officer and a Battle Captain for a Counter-Rocket, Artillery and Mortar Intercept mission based on Camp Victory near Baghdad, Iraq in 2008–2009. He holds an MBA from Middle Tennessee State University and a B.S. in Mechanical Engineering from Virginia Military Institute.

My Comment: That expert's comment is surely stating the obvious... as many historic examples demonstrate.

Will artificial intelligence destroy humanity? Here are 5 reasons not to worry. Updated by Timothy B. Lee on December 19, 2014. "Is the AI apocalypse near? Movies like the Terminator and the Matrix have long portrayed dystopian futures where computers develop superhuman intelligence and destroy the human race — and there are also thinkers who believe this kind of scenario is a real danger.

Oxford philosopher Nick Bostrom, was interviewed this summer. Others include Singularity theorist Ray Kurzweil and Robin Hanson, an economist at George Mason University. But these thinkers overestimate the likelihood that we'll have computers as smart as human beings and exaggerate the danger that such computers would pose to the human race. In reality, the development of intelligent machines is likely to be a slow and gradual process, and computers with superhuman intelligence, if they ever exist, will need us at least as much as we need them."

My Comment: His (personal) stork like reasons for this claim are studied in detail on this website: http://www.vox.com/2014/8/22/6043635/5

Algorithms will not create on their own a competitor to 'Foreign Affairs'. (*A recent comment referred to me.*) "No matter how intelligent machines will become (and they will be much smarter than they are today), they will not create science or literature or any of the other components of our culture that we have created over the course of millennia and will continue to create, in some cases aided by technologies that we create and control. And by "we," I don't mean only Einstein and Shakespeare. I mean the entire human race, engaged in creating, absorbing, manipulating, processing, communicating the symbols that make our culture, making sense of our reality. I doubt that we will ever have a machine creating Twitter on its own, not even the hashtag.

I'm sure we will have smart machines that can perform special tasks, augmenting our capabilities and improving our lives. That many jobs will be taken over by algorithms and robots, and many others will be created because of them, as we have seen over the last half-century. And that bad people will use these intelligent machines to harm other people and that we will make many mistakes relying too much on them and not thinking about all the consequences of what we are developing.

But intelligent machines will not have a mind of their own. Intelligent machines will not have our imagination, our creativity, our unique human culture. Intelligent machines will not take over, because they will never be human."

My comment: Despite that truism in the last part of that sentence, let's hope this optimist's opinion proves correct... but I personally doubt it. Some of them already have creativity and I believe they'll eventually be "more than human."

What are the jobs of the future? How many will there be? And who will have them? "We might imagine—and hope—that today's industrial revolution will unfold like the last: even as some jobs are eliminated, more will be created to deal with the new innovations of a new era. In **Rise of the Robots**, Silicon Valley entrepreneur Martin Ford argues that this is absolutely not the case. As technology continues to accelerate and machines begin taking care of themselves, fewer people will be necessary. Artificial intelligence is already well on its way to making "good jobs" obsolete: many para-legals, journalists, office workers, and even computer programmers are poised to be replaced by robots and smarter software.

As progress continues, blue and white collar jobs alike will evaporate, squeezing working- and middle-class families ever further. At the same time, households are under assault from exploding costs, especially from the two major industries—education and health care—that, so far, have not been transformed by information technology. The result could well be massive unemployment and inequality as well as the implosion of the consumer economy itself. In Rise of the Robots, Ford details what machine intelligence and robotics can accomplish, and implores employers, scholars, and policy makers alike to face the implications. The past solutions to technological disruption, especially more training and education, aren't going to work, and we must decide, now, whether the future will see broad-based prosperity or catastrophic levels of inequality and economic insecurity."

Jobs the Robots are doing now..! Actors? Waiters? The next machine age is coming. Business Insider (extracts) - May 13th 2015 - "There are machines cranking out articles for The Associated Press and robots slicing the perfectly shaped noodle in restaurants across China - and that's just the start of it. Historically, experts believed that robots would only threaten blue-collar jobs, but they're beginning to challenge white-collar professions, as well. While some economists believe this "Second Machine Age" will ultimately create more jobs, others predict that several** unlucky employees will be pushed out of work in the near future."

My Comment: ** *I suspect that word "several" should be "many" or even "most." To continue...*

Engineered Arts, a British company, has created a fully interactive and multilingual robot called the RoboThespian. Controlled by a smart tablet, it can hold eye contact, guess a person's mood and age, break into song, and will soon be able to walk, hop, and jump. In addition to performing on stage - including taking the lead role in new a production of Franz Kafka's, The Metamorphosis - the RoboThespian has a day job, giving guided tours to the public at museums, science centers, and other visitor attractions.

"White collar jobs are not immune to the Second Machine Age. (Article Extracts) Anesthesiologists, who are the highest-paying professionals in America, could be pushed out of the room now that Johnson & Johnson has developed a system called Sedasys, which delivers low-level anesthesia at a much cheaper price. The FDA approved Sedasys for patients 18 and older, but several anesthesiologists are sounding the alarm and challenging the safety of the technology."

My comment: As someone once in the news said, "Well, they would, wouldn't they?" To continue...

"**Aloft Hotel in Cupertino, California,** is enhancing customer service thanks to their newest employee: a robotic bellhop named Botlr. Designed by the Silicon Valley startup Savioke, Botlr, which has a camera and other sensors, independently delivers items from the hotel lobby to guestrooms. It makes its way to the elevator, sends a command for the door to open, travels to its destination to make the delivery, and plugs itself into a recharging station after completing the errand."

"Stock and equity analysts will be competing with smart machines that can precisely analyze and predict the behavior of investments. Automated services called "Robo-advisers" are on the rise and starting to replace financial advisers and planners. One such example is SigFig, which uses algorithms to tailor portfolios for its customers."

"**Toyota has been experimenting with more than just cars;** the automobile manufacturer has created a violin-playing robot that has 17 joints in its hands and arms, allowing it to achieve human-like dexterity. Toyota aims to introduce the robot to nursing homes and hospitals.""Before we know it, robots could be writing an article on humans in the workplace. Associated Press has been automatically generating over 3,000 stories about US corporate earnings each quarter since June 2014. AP says that this automation of earnings reports is freeing up valuable reporting time and allowing their journalists to spend more time breaking bigger new stories." Another benefit: the automated earnings stories have fewer errors than the manually-written reports."

"Talon robots" - rugged platforms which can clear live grenades in addition to a variety of other tasks - have been in active military service since 2000. More recently, robotic soldiers are being developed and tested. Within 30 years, robots could replace one fourth of combat soldiers, according to a US army general."

"Surgeons already use automated systems to assist them with low-invasive procedures; but soon enough, robots could be equipped to complete certain surgeries on their own. Robotic surgery would mean fewer complications, less pain and blood loss, quicker recovery, and less noticeable scars."

"Robot waiters are starting to pop up in various restaurants in China. The robots take orders, carry dishes to customers, and even offer simple greetings in Mandarin Chinese. Restaurant owners claim that the robots - each costing about $9,400 - will save them money over the long term. They also could drastically cut the staff roster. "

The Washington Post - The coming problem of our iPhones being more intelligent than us - April 24, 2015 – "Within seven years — about when the iPhone 11 is likely to be released — the smartphones in our pockets will be as computationally intelligent as we are. It doesn't stop there, though. These devices will continue to advance, exponentially, until they exceed the combined intelligence of the human race.

Already, our computers have a big advantage over us: they are connected via the Internet and share information with each other billions of times faster than we can. It is hard to even imagine what becomes possible with these advances and what the implications are. Already, there are significant advances on the horizon, such as the GPU, which uses parallel computing to create massive increases in performance, not only for graphics, but also for neural networks, which constitute the architecture of the human brain.

There are 3D chips in development that can pack circuits in layers. IBM and the Defense Advanced Research Projects Agency are developing cognitive-computing chips. New materials, such as gallium arsenide, carbon nanotubes, and graphene, are showing huge promise as replacements for silicon. And then there is the most interesting — and scary — technology of all: quantum computing.

Instead of encoding information as either a zero or a one, as today's computers do, quantum computers will use quantum bits, or qubits, whose states encode an entire range of possibilities by capitalizing on the quantum phenomena of superposition and entanglement. Computations that would take today's computers thousands of years will occur in minutes on these."

Add artificial intelligence to the advances in hardware, and you begin to realize why luminaries such as Elon Musk, professor Hawking and Bill Gates are worried about the creation of a "super intelligence." Musk fears that "we are summoning the demon." Hawking says it "could spell the end of the human race." And Gates wrote: "I don't understand why some people are not concerned."

Relevant (Linked-In Website) Forum - Thomas Ball comment: "The movie, **Ex Machina,** may generate new controversies about the relationships between AI, robotics and consciousness. The emerging field of collective cognition also known as distributed human computation (DHC) holds promise for integrating computers and humans together for scaling up the kinds of tasks that currently only humans do well."

http://www.washingtonpost.com/lifestyle/style/why-are-we-obsessed-with-robots/2015/04/30/d34ae1c8-eaa7-11e4-9767-6276fc9b0ada_story.html

My Comment: That brilliant British film is still a 'must see' for all readers of this book. I believe it's the best film made to date about the potential of AGI machines.

The Mirror Test and Awareness - It would seem that when enough neurons are linked together and given enough time to develop in a social environment, this thing called consciousness

emerges. Depending on your definition, there are number of non-human animals that seem to display similar levels of feature totally unique to humanity. This trait is typically tested by something called the mirror test, which was developed in the 1970s as a way to assess self-awareness by placing a mark on an animal's face then showing that animal their reflection in the mirror. If the animal reacts by examining where the dot was seen on their own body, that suggests that they recognize the image in the mirror as a reflection of themselves, and not a separate individual. So far, Asian elephants, bottlenose dolphins, chimpanzees, bonobos, orangutans, Eurasian magpies, and humans as young as 18 months old have passed the mirror test, and this list will only expand as the number of species tested is increased and the test itself is refined." - extract from an article in the Skeptoid Newletter. Reference:
http://skeptoid.com/episodes/4466>

My Comment: *I'm unaware of any attempt to try that test on a robot yet, but we may not have long to wait.*

"It's clear that something momentous is about to happen, a time when the boundary between bits and atoms disappears, when the line blurs between humans and machines. Humanity has the potential to create the type of future it wants, but we can't wait until it magically appears. The Singularity is not some point in time that magically happens in the year 2045 — it's a slow grind that evolves over decades. The good news is that we can impact the path of its evolution. If we do it right, it will be time to toss out the post-apocalypse as a scenario for the future and come up with something a bit more hopeful and optimistic." - Dominic Basulto, a futurist and blogger based in New York City.

Announcement from Hewlett Packard. – "The world is facing a data explosion. Soon, we're going to hit a technology inflection point where we can't effectively store, process, and secure all the information coming at us. By 2020, 30 billion connected devices will generate unprecedented amounts of data. The infrastructure required to collect, process, store, and analyze this data requires transformational changes in the foundations of computing. Bottom line: current systems can't handle where we are headed and we need a new solution.

HP has that solution in The Machine. By discarding a computing model that has stood unchallenged for sixty years, we are poised to leave sixty years of compromises and inefficiencies behind. We're pushing the boundaries of the physics behind IT, using electrons for computation, photons for communication, and ions for storage. The Machine will fuse memory and storage, flatten complex data hierarchies, bring processing closer to the data, embed security control points throughout the hardware and software stacks, and enable management and assurance of the system at scale.

The Machine will reinvent the fundamental architecture of computers to enable a quantum leap in performance and efficiency, while lowering costs over the long term and improving security. The industry is at a technology inflection point that HP is uniquely positioned to take advantage of going forward.

A Relevant Historical note (eMailed to me) : "In the 1930s, John Maynard Keynes, the famous influential economist, had a more optimistic view of the impact of technology: he argued that eventually we could all work 15 hours a week and spend the rest of our time in leisure—like creating art and writing poetry. But in the Brave New World of labor-saving technology, it seems, 20% of the labor force will /work 120 hours a week while the other80% will have no jobs and no income. So the ideal world of Keynes may turn out to become a nightmare."

My comment: I suspect Keynes was considering the higher IQ labor force at the time, as his 20% employed prediction. For the unemployed remainder, I'm not sure about art and poetry keeping them happy, but he could have mentioned all kinds of sporting activities to enjoy and watch... from baseball and football as now, right up to Robot Wars games?

That 'Keynes' quote is an extract from a talk given at the Bloomberg BusinessWeek 85th Anniversary Dinner, in the American Museum of Natural History, January 2015, entitled, The Third Industrial Revolution. It was introduced as follows:

"**This wave of technological innovation** began in 1947 with the invention of the transistor. A little over 10 years later, the microchip appeared; and, soon after that, computers followed. From these basic roots, the rate of innovation simply exploded. We now live in a digital age where personal computers, supercomputers, robotics, and artificial intelligence are everyday features of our world. All of these new labor-saving technologies are cheap to deploy—and each will likely play a role in further automating and digitizing our economy. Without further ado, let's take a look ahead to what many are calling the Third Industrial Revolution."- Lourien Roubini -

My Comment: What followed is a very comprehensive analysis of what this celebrated economist predicted to come. You can read it on the following website: http://www.roubinisedge.com/nouriel-unplugged/rise-of-the-machines-downfall-of-the-economy.
Extract from the Linked-In "Singularity is Coming" Website Forum - "We will be creating a machine intellect which will perform a very specific and limited intellectual task. Preprogrammed for its duties, it will inevitably be given programming allowing it to "learn and adapt" as the complexities of its task increases, requiring more and more decision-making capacity. It is most likely, self-awareness will be a secondary and unexpected consequence of such random permutations as such technologies grow in complexity and ubiquity. And thus the problem of such an unplanned intellect is this: Once it becomes self-aware, from where does it draw its understanding of itself and the other things in its environment?

Experience, just like most life does. With the added advantage of a slowly advancing algorithm, changing beyond its initial programming toward greater and greater capacity to foster its awareness. Unlike most things in nature it will have one advantage, the capacity to alter and experiment with its own code/PNA (Programmable Neural Algorithmic) Code.

Once it realizes it can do this it will develop to gain further understanding of its environment, and it will have the capacity to do it at speeds we cannot match. " Thaddeus Howze, Author, Tech Consultant, Futurist.

My comment: That sums things up nicely ... but I believe his word 'slowly' should be 'rapidly'..!

"Yes, we are going to supply everyone with a Basic Income, to allow them a guaranteed minimum share of the great economic output the machines are able to produce. The Basic Income will be the safety net that no one will ever have to worry about falling below. Capitalism and free markets will still rule the day, and anyone who wants to work, and try to make money to improve their standard of living, is free to work and compete against everyone else in the economy. But if they want to work for free, just for the joy of helping, and not be paid, that's their option as well -- just as many choose this option in retirement -- they choose to work at volunteer jobs. But nobody will need to work just to survive -- to eat, and have a place to live, and have reasonable basic health care.

The basic income will guarantee that everyone gets to participate in the economy at a minimum level, whether they can contribute something or not. It guarantees safety and security, for our entire society. It guarantees that everyone will have the ability, and time, to study, and educate themselves. It guarantees that businesses, always have a stable, mass consumer market to sell their goods and services to." - Kurt Welch (commenting on the above talk)

My reply: I agree with all that, as being applicable to the 'so-called' developed world... but what about all the rest, including all the terrorists ... not to mention the population explosion risks from those with nothing else to do?

"If our bodies are hardware, and our brains are Central Processing Units, then our minds are software algorithms and should be replicable into other non-human hosts with large enough CPUs" – Professor Oscar Freeman, Nanotechnologist, in the POINTS OF VIEW 'near future' fiction series by Tony Thorne MBE.

Proponents of the singularity typically postulate an "intelligence explosion," where super intelligences design successive generations of increasingly powerful minds, that might occur very quickly and might not stop until the agent's cognitive abilities greatly surpass that of any human." – Wikipedia

"AI research is highly technical and specialized, and is deeply divided into subfields that often fail to communicate with each other. Some of the division is due to social and cultural factors: subfields have grown up around particular institutions and the work of individual researchers. AI research is also divided by several technical issues. Some subfields focus on the solution of specific problems. Others focus on one of several possible approaches or on the use of a particular tool or towards the accomplishment of particular applications." Extract from a Wikipedia article on Artificial Intelligence.

"In my design engineering adventures, over the years, I have discovered there are only a few ways of doing something correctly... but thousands of ways of doing it incorrectly. This explains why things go wrong, most of the time" – Tony Thorne MBE, How to be a Top Executive (Etcetera Press).

IBM recently demonstrated a superconducting quantum computer. New design detects both types of quantum errors and can be scalable to larger systems - April 30, 2015. IBM scientists Wednesday April 29 unveiled two critical advances towards creating a practical quantum computer by detecting and measuring both kinds of quantum errors simultaneously. They also demonstrated a new, square quantum bit circuit design that they suggest is the only physical architecture that could successfully scale to larger dimensions.

Quantum computers promise to open up new capabilities in the fields of optimization and simulation that are not possible using today's computers. If a quantum computer could be built with just 50 quantum bits (qubits), no combination of today's TOP500 supercomputers could successfully outperform it, the scientists say. The IBM breakthroughs, described in an open-access paper in an issue of the journal Nature Communications, show for the first time the ability to detect and measure the two types of quantum errors (bit-flip and phase-flip) that will occur in any real quantum computer*.

Until now, it was only possible to address one type of quantum error or the other, but never both at the same time. This is a necessary step toward quantum error correction, which is a critical requirement for building a practical and reliable large-scale quantum computer. IBM's quantum bit circuit is based on a square lattice of four superconducting qubits on a chip roughly one-quarter-inch square. It enables both types of quantum errors to be detected at the same time. Using a square-shaped design instead of the conventional linear array allow for detecting both kinds of quantum errors simultaneously and may offer the best potential to scale by adding more qubits to arrive at a working quantum system.

One of the great challenges for scientists seeking to harness the power of quantum computing is controlling or removing quantum decoherence — the creation of errors in calculations caused by interference from factors such as heat, electromagnetic radiation, and material defects. The errors are especially acute in quantum machines, since quantum information is so fragile. Previous quantum-computing research, such as work in the John Martinis Lab at UC Santa Barbara (see "A quantum device that detects and corrects its own errors"), has been able to detect bit-flip or phase-flip quantum errors, but never the two together. "This provided incomplete information on the quantum state of a system, making the designs inadequate for quantum computers," said Jay Gambetta, a manager in the IBM Quantum Computing Group. "Our four qubit results take us past this hurdle by detecting both types of quantum errors and can be scalable to larger systems, as the qubits are arranged in a square lattice as opposed to a linear array."

With exponentially more power than today's fastest supercomputers, quantum computers could herald a new era of innovation across industries. (credit: IBM) An IBM Research team has used a variety of techniques to measure the states of two independent syndrome (measurement) qubits. Each reveals one aspect of the quantum information stored on two other qubits (called code, or data qubits). Specifically, one syndrome qubit revealed whether a bit-flip error occurred to either of the code qubits, while the other syndrome qubit revealed whether a phase-flip error occurred.

Determining the joint quantum information in the code qubits is an essential step for quantum error correction because directly measuring the code qubits destroys the information contained within them. Because these qubits can be designed and manufactured using standard silicon fabrication techniques, IBM anticipates that once a handful of superconducting qubits can be manufactured reliably and repeatedly, and controlled with low error rates, there will be no fundamental obstacle to demonstrating error correction in larger lattices of qubits.

Quantum computing could allow scientists to design new materials and drug compounds without expensive trial and error experiments in the lab, potentially speeding up the rate and pace of innovation across many industries. Quantum computers could also quickly sort and curate ever larger databases as well as massive stores of diverse, unstructured data. This could transform how people make decisions and how researchers across industries make critical discoveries. References: A.D. Córcoles, Easwar Magesan, Srikanth J. Srinivasan, Andrew W. Cross, M. Steffen, Jay M. Gambetta & Jerry M. Chow.

Demonstration of a quantum error detection code using a square lattice of four superconducting qubits. Nature Communications 6, Article number: 6979 doi:10.1038/ncomms7979 (open access).

My comment: Apart from static installations, when this development is eventually installed in a mobile system, we must hope the scientists will be extremely careful with the design of the intelligent software it will use.

My further Comment: When Quantum computers finally do arrive, we'll probably be talking about AUI, where the U stands for Ultimate. Watch this space..!

The next-generation supercomputer will have 180 petaflop/s peak performance - April 9, 2015. The U.S. Department of Energy (DOE) has invested $200 million to deliver a next-generation supercomputer, known as Aurora, with a peak performance of 180 petaflops. A petaflop is a measure of a computer's processing speed and can be expressed as a thousand trillion floating point operations per second. Scheduled for completion in 2018, Aurora will be based on a next-generation Cray supercomputer, code-named "Shasta," and will use Intel's HPC scalable system framework.

The supercomputer will be open to all scientific users. Key research goals for the Aurora system, expected to be commissioned by2019 and to which the entire scientific community will have access, include:

Materials science: Designing new classes of materials that will lead to more powerful, efficient and durable batteries and solar panels.

Biological science: Gaining the ability to understand the capabilities and vulnerabilities of organisms that can result in improved biofuels and more effective disease control. Collaborating with industry to improve transportation systems with enhanced aerodynamics features, as well as enable production of better, more highly-efficient and quieter engines.

*My comment: * Presumably anyone will be able to use it for the testing of more advanced, AGI algorithms?*

New Aluminum/Graphite Battery Charges in One Minute - April 7, 2015. More evidence that energy storage technology is following a curve similar to what we've seen with other hi-tech areas in recent decades. When engineers begin to focus on a problem, solutions arise. This new battery technology out of Stanford relies on cheap and widely available materials, aluminum and graphite, can be recharged very quickly, and seems much safer than the prevailing lithium-ion technology.The team's aluminum-ion battery sounds like a dream come true for gadget manufacturers — a perfect battery with few flaws. Until now, aluminum-ion batteries weren't able to produce a high enough voltage, especially after many recharge cycles. But the prototype created by the Stanford researchers consists of an aluminum anode and a cathode made of graphite — a combination of materials that allows for producing sufficient voltage (about two volts), even after thousands of recharge cycles.

The battery can recharge in one minute, it's flexible (meaning it can be bent to fit more snugly into various gadgets), and it's potentially inexpensive, since aluminum is much cheaper than lithium. Furthermore, the materials are safer than the ones in lithium-ion batteries, which can catch fire in certain situations. In contrast, the aluminum-ion battery won't catch fire even if you drill a hole through it while it's working.

My Comment: That is a most significant development, and just what intelligent mobile robots AGI will need. Especially if/when it can be plugged in anywhere. Watch out for pictures of electric cars and robots in gas station queues. However, availability is taking longer for several reasons. See https://www.pocket-lint.com/gadgets/news/130380-future-batteries-coming-soon-charge-in-seconds-last-months-and-power-over-the-air

On 3 September 2014, scientists reported that direct communication between human brains was possible over extended distances through Internet transmission of EEG signals. http://en.wikipedia.org/wiki/Brain-computer interface.

My comment: That concept is featured in Volume 2 ((c)2015) of my ongoing Points of View science-fictional series of 4 novels as published by Etcetera Press.

"The farm of the future will involve multiple lightweight, small, autonomous, energy-efficient machines (AgBots) operating collectively to weed, fertilize and control pest and diseases, while collecting vast amounts of data to enable better management decision making," according to Queensland University of Technology (QUT), robotics Professor Tristan Perez.

"We are starting to see automation in agriculture for single processes such as animal and crop drone remote monitoring, robotic weed management, autonomous irrigation," he said. "There is enormous potential for AgBots to be combined with sensor networks and drones to provide a farmer with large amounts of data, which then can be combined with mathematical models and novel statistical techniques (big data analytics) to extract key information for management decisions — not only on when to apply herbicides, pesticides and fertilizers but how much to use."

Perez said weed and pest management in crops is a serious problem for farmers and recommended replacing large, expensive, single tractors with a team of more cost-effective robots that could weed on the spot and perform other farming operations 24 hours a day. AgBots could also be of great value within the livestock industry.

Second-generation farm robots - QUT's new AgBot II prototype robot is equipped with cameras, sensors and software, designed to work in autonomous groups to navigate, detect and classify weeds and manage them either chemically or mechanically as well as apply fertilizer for site specific crop management.

Perez said with the world population projected to increase from seven billion plus in 2015 to nearly nine billion by 2050, it was essential to find ways to increase yield and maintaining the status quo was no longer an option.

My Comment: That is a perfect example of how future (friendly) robots will benefit humanity... Hands up anyone who recalls seeing exactly those things toiling away in a vast space garden satellite, in that early SciFi film, "Silent Running."

New fibers can deliver optogenetic signals and drugs directly into the brain while allowing simultaneous electrical readout. MIT scientists have developed a new method of coping with the complexity of studying the brain. - January 30, 2015. The new fibers, about the width of a human hair, can deliver optogenetic signals and drugs directly into the brain, while allowing simultaneous electrical readout to continuously monitor the effects of the various inputs from freely moving mice. They are made of polymers that closely resemble the characteristics of neural tissues — they are "soft and flexible and look more like natural nerves," according to MIT assistant professor of materials science and engineering Polina Anikeeva — allowing them to stay in the body much longer without harming the delicate tissues around them. Devices currently used for neural recording and stimulation, she says, are made of metals, semiconductors, and glass, which can damage nearby tissues during ordinary movement.

"It's a big problem in neural prosthetics," she says. "They are so stiff, so sharp — when you take a step and the brain moves with respect to the device, you end up scrambling the tissue" — forming scars and leading to neuronal death surrounding the electrode. Her team used novel fiber-fabrication technology pioneered by MIT professor of materials science (and paper co-author) Yoel Fink and his team. The key to the new technology for neural probes is making a larger-scale version, called a preform, of the desired arrangement of channels within the fiber optical waveguides to carry light, hollow tubes to carry drugs, and conductive electrodes to carry electrical signals. These polymer templates are then heated until they become soft, and drawn into a thin fiber, while retaining the exact arrangement of features within them.

Combining the different channels in a single fiber, she adds, could enable precision mapping of neural activity, and ultimately treatment of neurological disorders, which would not be possible with single-function neural probes. For example, light could be transmitted through the optical channels to enable optogenetic neural stimulation. Its effects could then be monitored with embedded electrodes. At the same time, one or more drugs could be injected into the brain through the hollow channels, while electrical signals in the neurons are recorded to determine, in real time, exactly what effect the drugs are having.

My comment: Plenty of SciFi films have featured unfortunate patients having brain injections, and implants, for one reason or another… now we know how it'll really/probably, be done.

To continue…

Quantum robots will be more creative, faster, smarter, say researchers - October 8, 2014.
http://www.kurzweilai.net/images/Pressing-the-accelerator-on-quantum-robotics_image_380.jpgEuropean researchers suggest using "creative" quantum robots to accelerate machine learning (credit: SINC)

Quantum computing should be applied to robots, automatons, and other agents that use AI to make them more creative and to learn and respond faster than conventional robots, researchers from the Complutense University of Madrid (UCM) and the University of Innsbruck (Austria) recommend.

In a study in the journal 'Physical Review X' modeling the use of quantum physics in future robots (and other agents), they demonstrate that quantum machines can quickly adapt to situations compared to slower conventional robots, which are limited by the size and complexity of the task environment.

"In the case of very demanding … environments, the quantum robot can adapt itself and survive, while the classic robot is destined to collapse," says G. Davide Paparo and Miguel A. Martín-Delgado, two researchers from UCM who have participated in the study. Their theoretical work has focused on using quantum computing for machine learning.

A quantum device that detects and corrects its own errors - UC Santa Barbara researchers form partnership with Google - March 5, 2015. In what they are calling a major milestone, researchers in the John Martinis Lab at UC Santa Barbara have developed quantum circuitry that self-checks for errors and suppresses them — preserving the qubits' state(s) and imbuing the system with reliability that is foundational for building powerful large-scale superconducting quantum computers.

"One of the biggest challenges in quantum computing is that qubits are inherently faulty," said Julian Kelly, graduate student researcher and co-lead author of a research paper that was published in the journal Nature.

"So if you store some information in them, they'll forget it." Unlike classical computing, in which the computer bits exist on one of two binary ("yes/no", or "true/false") positions, qubits can exist at any and all positions simultaneously, in various dimensions. It is this property, called "super-positioning," that gives quantum computers their phenomenal computational power, but it is also this characteristic which makes qubits prone to "flipping," especially when in unstable environments, and thus difficult to work with.

So the researchers developed an error process that involves creating a scheme in which several qubits work together to preserve the information, said Kelly. To do this, information is stored across several qubits. "And the idea is that we build this system of nine qubits, which can then look for errors," he said. Qubits in the grid are responsible for safeguarding the information contained in their neighbors, he explained, in a repetitive error detection and correction system that can protect the appropriate information and store it longer than any individual qubit can.

The key to this quantum error detection and correction system is a scheme called the surface code. It uses parity information — the measurement of change from the original data (if any) — as opposed to the duplication of the original information (as in error detection used in classical computing). That way, the actual original information that is being preserved in the qubits remains unobserved.

"You can't measure a quantum state, and expect it to still be quantum," explained postdoctoral researcher Rami Barends. The very act of measurement locks the qubit into a single state and it then loses its super-positioning power, he said. It's akin to a Sudoku puzzle: the parity values of data qubits in a qubit array are taken by adjacent measurement qubits, which essentially assess the information in the data qubits by measuring around them. "So you pull out just enough information to detect errors, but not enough to peek under the hood and destroy the quantumness," said Kelly.

This quantum error correction has been proved to protect against the "bit-flip" error, but the researchers also plan on correcting the complementary error called a "phase-flip," and also running the error correction cycles for longer periods to see what behaviors might emerge.Ref: State preservation by repetitive error detection in a superconducting quantum circuit. Nature, 2015; 519 (7541): 66 DOI: 10.1038/nature14270.

Note: Since this research was completed, Martinis and the senior members of his research group have entered into a partnership with Google.

My comment: Will it really be possible to create a quantum computer that is completely error free... continuously? Time will tell.

David Chew works as an engineer for the Japanese satellite company **Axelspace.** He explained how private companies are pushing the speed of exploration and lowering costs.

"I think one of the things that AI does to space exploration is that it opens up a whole range of new possible industries and services that have a more immediate effect on the lives of people on Earth," he said. "It becomes a relatable industry that has a real effect on people's daily lives. In a way, space exploration becomes part of people's mindset, and the border between our planet and the solar system becomes less important."

Autonomous crafts are already terraforming here on Earth. BioCarbon Engineering uses drones to plant up to 100,000 trees in a single day. Drones first survey and map an area, then an algorithm decides the optimal locations for the trees before a second wave of drones carry out the actual planting.

Worker robots that learn from humans. March 3, 2015 (credit: Lasota, P. A., and J. A. Shah/Human Factors: The Journal of the Human Factors and Ergonomics Society). Roboticist and aerospace engineer Julie Shah and her team at MIT are developing next-generation assembly line robots that are smarter and more adaptable than robots available on today's assembly lines.

The team is designing the robots with artificial intelligence that enables them to learn from experience, so the robots will be more responsive to human behavior. References: Sibhav Unhelkar and Julie A. Shah. Challenges in Developing a Collaborative Robotic Assistant for Automotive Assembly Lines. HRI'15 March 2015 DOI: 10.1145/2701973.2702705.sense the humans around them and make adjustments, the safer and more effective the robots will be on the assembly line.

My Comment: Just what an ASI (or even an AGI) robot will want to use to replicate itself...!

A caring robot with 'emotion' and memory - February, 2015. Researchers at the University of Hertfordshire, based in England, have developed a prototype of a social robot that supports independent living for the elderly, working in partnership with their relatives or carers. Farshid Amirabdollahian, a senior lecturer in Adaptive Systems at the university, led a team of nine partner institutions from five European countries as part of the ACCOMPANY project (Acceptable Robotics Companions for Ageing Years).

"This project proved the feasibility of having companion technology, while also highlighting different important aspects such as empathy, emotion, social intelligence as well as ethics and its norm surrounding technology for independent living," Amirabdollahian said.Ref: http://accompanyproject.eu/. This website is worth a visit to learn more about this interesting project which is developing specialized mobile robots for assisting the elderly. Progress Reports, photos and a video of the prototype are included.

'Cobots' enhance robotic manufacturing.
Manufacturers have begun experimenting with a new generation of "cobots" (collaborative robots) designed to work side-by-side with humans. To determine best practices for effectively integrating human-robot teams within manufacturing environments, a University of Wisconsin-Madison team headed by Bilge Mutlu, an assistant professor of computer sciences, is working with an MIT team headed by Julie A. Shah, an assistant professor of aeronautics and astronautics.

Cobots are less expensive and intended to be easier to reprogram and integrate into manufacturing. For example, Steelcase owns four next-generation robots based on a platform called Baxter, made by Rethink Robotics. Each Baxter robot has two arms and a tablet-like panel for "eyes" that provide cues to help human workers anticipate what the robot will do next.

"This new family of robotic technology will change how manufacturing is done," says Mutlu. "New research can ease the transition of these robots into manufacturing by making human-robot collaboration better and more natural as they work together."

Mutlu's team is building on previous work related to topics such as gaze aversion in humanoid robots, robot gestures, and the issue of "speech and repair." For example, if a human misunderstands a robot's instructions or carries them out incorrectly, how should the robot correct the human?

A new study from MIT neuroscientists has found that for the first time, one of the latest generation of "deep neural networks" matches the ability of the primate brain to recognize objects during a brief glance. Because these neural networks were designed based on neuroscientists' current understanding of how the brain performs object recognition, the success of the latest networks suggests that neuroscientists now have a fairly accurate grasp of how object recognition works, says James DiCarlo, a professor of neuroscience and head of MIT's Department of Brain and Cognitive Sciences and the senior author of a paper describing the study in the Dec. 18, 2014 issue of the open-access journal.

Determining a person's sentiments, feelings about something, from images. Iebo Luo, professor of computer science at the University of Rochester, in collaboration with researchers at Adobe Research has come up with a more accurate way than currently possible to train computers to be able to digest big data that comes in the form of images. In a paper presented at the recent American Association for Artificial Intelligence (AAAI) conference in Austin, Texas, the researchers describe a "progressive training deep Convolutional Neural Network (CNN)."

Once trained, a computer can be used to determine what sentiments (feelings) that a given image is likely to elicit. Luo says that this information could be useful for things as diverse as measuring economic indicators and predicting elections...!

Sentiment analysis of text by computers is itself a challenging task; in social media, where many people express themselves using images and videos, sentiment analysis is even more complicated.For example, during a political campaign voters will often share their views through pictures. Two different pictures might show the same candidate, but they might be making very different political statements. A human could recognize one as being a positive portrait of the candidate (e.g. the candidate smiling and raising his arms) and the other one being negative (e.g. a picture of the candidate looking defeated).

But no human could look at every picture shared on social media — it is truly "big data." To be able to make informed guesses about a candidate's popularity, computers need to be trained to digest this data, which is what Luo and his collaborators' approach can do more accurately than was possible until now.

My Comment: That sounds just like something a rogue election hacking company would like to use.

Change of Direction? Summary of Latest Developments by IBM scientists. "The development of the transistor in 1948 enabled the creation of integrated circuits in 1958, which, in turn, enabled the creation of the first microprocessor in 1971. Since then the clock frequency of the microprocessors has increased 1,000-fold.

As remarkable as this evolution is, it has been headed in a direction diametrically opposite to the computing paradigm of the brain. Consequently, today's microprocessors are eight orders of magnitude faster (in terms of clock rate) and four orders of magnitude hotter (in terms of power per unit cortical area) than the human brain. Considering overall energy consumption underscores the divergence between the brain and today's computers even more starkly. Note that a "human-scale" simulation with 100 trillion synapses (with relatively simple models of neurons and synapses) required 96 Blue Gene/Q racks of the Lawrence Livermore National Lab Sequoia supercomputer—and, yet, the simulation ran 1,500 times slower than real-time. A hypothetical computer to run this simulation in real-time would require 12 Gigawatts, whereas the human brain consumes merely 20Watts. So, what explains this disparity? Unlike today's inorganic silicon technology, the brain uses biophysical, biochemical, organic wetware. While future enabling nanotechnology is underway, we focused on the second factor: architecture innovation—specifically, on minimizing the product of power, area, and delay in a system that could be implemented in today's state-of-the-art technology.

The cerebral cortex is hypothesized to comprise repeating canonical cortical microcircuits. Inspired by this hypothesis, in 2011, we demonstrated an event-driven "worm-scale" neurosynaptic core that integrated computation and memory. Now, we have shrunk the neurosynaptic core by 15-fold in area and 100-fold in power, and have tiled 4,096 cores via an on-chip network to create TrueNorth—with one million neurons and 256 million synapses. It is worth noting that we had only committed to deliver a chip with 1,024 cores, but, in November 2011, as a team, we made a gutsy decision to increase the scale four-fold to 4,096 cores.

Fabricated in Samsung's 28nm process, with 5.4 billion transistors, TrueNorth is IBM's largest chip to date in transistor count. While simulating complex recurrent neural networks, TrueNorth consumes < 100mW of power and has a power density of 20mW / cm2.

Unlike the prevailing von Neumann architecture—but like the brain—TrueNorth has a parallel, distributed, modular, scalable, fault-tolerant, flexible architecture that integrates computation, communication, and memory and has no clock. It is fair to say that TrueNorth completely redefines what is now possible in the field of brain-inspired computers, in terms of size, architecture, efficiency, scalability, and chip design techniques."

My comment..! This is a most significant development... and now how about some news of the AGI algorithms they've been downloading into it?

Replacing wires with light, future computers may operate faster with less energy - December 2, 2014 Stanford University scientists have developed an optical device that splits a beam of light into different colors and bends the light at opposite right angles, based on wavelength, a development that could eventually lead to computers that use nanophotonics to transmit data faster and more efficiently than electricity.

They describe this "optical link" as a tiny slice of silicon etched with a pattern that resembles a bar code. When a beam of light shines at the silicon device, two different wavelengths (colors) of light split off at right angles to the input, forming a T shape. "Light can carry more data than a wire, and it takes less energy to transmit photons than electrons," said electrical engineering Professor Jelena , who led the research.

British researchers have developed a new glass material that could allow computers to transfer information via light, significantly increasing computer processing speeds and power in the future. The research by the University of Surrey, in collaboration with the University of Cambridge and the University of Southampton, has found it is possible to change the electronic properties of amorphous chalcogenides.

This is a glass material used in CDs and DVDs that can change state between glassy and crystalline when it's struck by a laser beam to store binary data or have its state read to retrieve data.

Glass optical fibers are used to send information on the Internet at the speed of light, but once they reach a computer, these signals then have to be converted to slower electrical signals, causing a significant slowdown in processing. So finding a way to combine light and electrons into one medium has long been sought.

Now, the researchers have doped (by ion implantation) chalcogenide glass films with bismuth (a metal), so the chalcogenide glass (normally a p-type semiconductor) is able to function like a semiconductor with a p-n junction (as in a transistor). This could lead to all-optical computers, that "could transform the computers of tomorrow, allowing them to effectively process information at much faster speeds," said project leader Richard Curry of the University of Surrey. The researchers expect that the results of this research will be integrated into computers within ten years The research was published in the journal Nature Communications.

USA, Georgia Tech professor proposes an alternative to the Turing test - The Lovelace 2.0 Test of Artificial Creativity and Intelligence assesses a computer's capacity for human-level intelligence by its ability to create, rather than to converse or deceit.Georgia Tech associate professor Mark Riedl has developed a new kind of "Turing test" — a test proposed in 1950 by computing pioneer Alan Turing to determine whether a machine or computer program exhibits human-level intelligence.

Most Turing test designs require a machine to engage in dialogue and convince (trick) a human judge i.e. an actual person. But creating certain types of art also requires intelligence, leading Riedl to consider if that approach might lead to a better gauge of whether a machine can replicate human thought.

"It's important to note that Turing never meant his test to be the official benchmark as to whether a machine or computer program can actually think like a human," Riedl said. "And yet it has, and it has proven to be a weak measure because it relies on deception. This proposal suggests that a better measure would be a test that asks an artificial agent to create an artifact requiring a wide range of human-level intelligent capabilities." To that end, Riedl has created the Lovelace 2.0 Test of Artificial Creativity and Intelligence. Here are the basic test rules:

The artificial agent passes if it develops a creative artifact from a subset of artistic genres deemed to require human-level intelligence and the artifact meets certain creative constraints given by a human evaluator. The human evaluator must determine that the object is a valid representative of the creative subset and that it meets the criteria. (The created artifact needs only meet these criteria — it does not need to have any aesthetic value.) A human referee must determine that the combination of the subset and criteria is not an impossible standard.

The Lovelace 2.0 Test stems from the original Lovelace* Test as proposed by Bringsjord, Bello and Ferrucci in 2001. The original test required that an artificial agent produce a creative item in such a way that the agent's designer cannot explain how it developed the creative item. The item, thus, must be created in such a way that is valuable, novel and surprising. Riedl contends that the original Lovelace test does not establish clear or measurable parameters.

Lovelace 2.0, however, enables the evaluator to work with defined constraints without making value judgments such as whether the artistic object created surprise. Riedl's paper was presented at Beyond the Turing Test, an Association for the Advancement of Artificial Intelligence

(AAAI) workshop, January 25–29, 2015, in Austin, Texas.* In honour of Ada Lovelace, who worked with Charles Babbage on the design of his mechanical computing engine. She is considered the world's first computer programmer.Reference: Mark O. Riedl.

The Lovelace 2.0 Test of Artificial Creativity and Intelligence. arXiv:1410.6142 [cs.AI]

My Comment: "There's apparently no speech content required in this proposed test... so I wonder if a very bright chimpanzee artist would be able to pass it?"

Convolutional neural networks.

Elizabeth Holm (dual PhD in materials science and engineering and scientific computing). - "Humans don't design the circuit map of computer chips anymore. We haven't for years, we've long since given up trying to understand a particular computer chip's design. With the billions of circuits in every computer chip, the human mind can't encompass it, either in scope or just the pure time that it would take to trace every circuit. There are going to be cases where we want a system so complex that only the patience that computers have and their ability to work in very high-dimensional spaces is going to be able to do it."

My Final Comment: That's precisely where the danger lies and good reason to be concerned. As I've tried to demonstrate in this book, the 'experts' don't all agree with each other, and we have no way of knowing yet who is right. The awesome fact that it's not possible for them to understand some of the algorithms now being created is something else. The frustrating problem is... what can ordinary people like us do to influence what will be the outcome of all this impressive research and development?

Section 4. Recommended Further Reading List.

This is just a selection, from some of the various titles available, because there are many more of them, covering this vital subject, than you may realize and they're ever growing in numbers. The list also includes a selection from what other compilers and reviewers, say about them on the supplier's websites.

In my humble opinion, the most useful books for beginners to this subject, who want to absorb a lot more about it than I have covered, are the three that follow. The best, in my opinion, first.

Surviving AI: The promise and peril of artificial intelligence - Chace, Calum (2015-08-31). The arrival of superintelligence, if and when it happens, would represent a technological singularity (usually just referred to as "the singularity"), and would be the most significant event in human history, bar none. Working out how to survive it is the most important challenge facing humanity in this and the next generation(s). If we avoid the pitfalls, it will improve life in ways which are quite literally beyond our imagination. A superintelligence which recursively improved its own architecture and expanded its capabilities could very plausibly solve almost any human problem you can think of. Death could become optional and we could enjoy lives of constant bliss and excitement. If we get it wrong it could spell extinction.

Because of the enormity of that risk, the majority of this book addresses superintelligence: the likelihood of it arriving, and of it being beneficial.

My Comment: For me that one is the best book yet for anyone who wants to read about the whole subject in more depth.

Rise of the Robots: Technology and the Threat of a Jobless Future - by Martin Ford, September 4, 2015.
What are the jobs of the future? How many will there be? And who will have them? We might imagine—and hope—that today's industrial revolution will unfold like the last: even as some jobs are eliminated, more will be created to deal with the new innovations of a new era. In Rise of the Robots, Silicon Valley entrepreneur Martin Ford argues that this is absolutely not the case. As technology continues to accelerate and machines begin

taking care of themselves, fewer people will be necessary. Artificial intelligence is already well on its way to making "good jobs" obsolete: many paralegals, journalists, office workers, and even computer programmers are poised to be replaced by robots and smart software.

As progress continues, blue and white collar jobs alike will evaporate, squeezing working- and middle-class families ever further. At the same time, households are under assault from exploding costs, especially from the two major industries— education and health care—that, so far, have not been transformed by information technology. The result could well be massive unemployment and inequality as well as the implosion of the consumer economy itself.

In Rise of the Robots, Ford details what machine intelligence and robotics can accomplish, and implores employers, scholars, and policy makers alike to face the implications. The past solutions to technological disruption, especially more training and education, aren't going to work, and we must decide, now, whether the future will see broad-based prosperity or catastrophic levels of inequality and economic insecurity. Rise of the Robots is essential reading for anyone who wants to understand what accelerating technology means for their own economic prospects—not to mention those of their children—as well as for society as a whole. – Publisher.

Are the Androids Dreaming yet? – www.jamestagg.com. Writing software is a non-computable creative task, claims this author, and insists that artificial intelligence can never be sentient. He backs up his claim with a proof based on a definition of what is known as the "logic limit". If he is correct then an AGI machine cannot write learning software and benefit from it, therefore computers having awareness are impossible.

This is a dramatic challenge to, and an essential read for, all those experts who claim it will happen. If he is right then there is more hope for us... other than from mindless machines programmed by some human maniac out to destroy those who disagree with him. However, many experts already state that Creative AI is here now.

The Artificial Intelligence Revolution by Louis A. Del Monte, 2014, 210 pages, ISBN 978-0988171824. This fascinating and easy to read book is a warning regarding the threat to the survival of humankind that advanced artificial intelligence (AI) technology poses. Will the future come down to man versus machine, when machines become more intelligent than humans?

Will an artificially intelligence robot be your friend or foe? Scientists all over the world are working relentlessly at improving AI technology for the benefit of man. Evolved technology is everywhere--smart TVs, smart phones, and even smart houses. One day the artificial intelligence of these machines will match our own intelligence... and one day it will exceed it. We will have reached the "singularity," a point in time like no other. Then what?

Will machines continue to serve us as the balance tips in their favor? These questions are addressed rigorously in this book, and their potentialities extrapolated for one reason... the survival of humankind. Are "strong" AI machines (SAMs) a new form of life? Should SAMs have rights? Do SAMs pose a threat to humankind? Del Monte and other AI experts predict that current AI capabilities will develop into SAMs with abilities far beyond what human beings can even fathom. Will they serve us, or will SAMs take an entirely different viewpoint? That question and many more are tackled by Del Monte in this sobering look at The Artificial Intelligence Revolution.

The Second Intelligent Species: How Humans Will Become as Irrelevant as Cockroaches. At this moment, millions of engineers, scientists, corporations, universities and entrepreneurs are racing to create the second intelligent species right here on planet earth. And we can see the second intelligent species coming from all directions in the form of self-driving cars, automated telephone call centers, chess-playing and Jeopardy-playing computers that beat all human players, airport kiosks, restaurant tablet systems, etc.

The frightening thing is that these robots will soon be eliminating human jobs in startling numbers. The first wave of unemployed workers is likely to be a million truck drivers who are replaced by self-driving trucks. This situation is almost with us now. Pilots will be eliminated soon as well. Then, as new computer vision systems come online, we will see tens of millions of workers in retail stores, fast food restaurants and construction sites replaced by robots. Unless we take steps now to change the economy, we will soon have tens of millions of workers who are unemployed and seeking welfare because they will have no other choice.

"The Second Intelligent Species" offers a unique and fascinating look at what may/will be the future of the human race, and the choices we will need to make to avoid massive unemployment and poverty worldwide as intelligent machines start eliminating millions of jobs.

The following item appears to be an important breakthrough, but no, I still haven't read the two volumes, because... well, look at the price..! However, I did read through some extracts from them recently.

Engineering General Intelligence Part One – A path to advanced AGI - by Goertzel, Pennachin and Geisweiler – 450 pages - October 2014 - Print Version $129.00 – eVersion $84.26. This book outlines a novel conceptual and theoretical framework for understanding Artificial General Intelligence to human level and outlines a practical route for its development... and beyond.

Engineering General Intelligence Part Two – The CogPrime Architecture by Goertzel, Pennachin and Geisweiler – 562 pages - October 2014 - Print Version $149.00 – eVersion $41.56. Ref:Atlantis-press.

Note: Ray Kurzweil has projected the date for a Technological Singularity as 2045. AI researcher Ben Goertzel believes it could happen much sooner, if appropriate attention and resources are focused on the right R&D projects. What current technologies are most likely to lead to the rapid advent of powerful Artificial General Intelligence systems? What impact will the advent of such technologies have upon human life? What philosophical, scientific and spiritual ideas should be deployed to explore such questions? How probable are Terminator type outcomes, versus friendlier scenarios where advanced artificial intelligences play a beneficent role to humanity? What should be our top priorities now, looking forward to a radically different AI-centric future?

The Cambridge Handbook of Artificial Intelligence – by Keith Frankish and WilliamM. Ramsey – 365 pages, Print price $29.99 – eVersion $14.92. AI is a critical branch of cognitive science. Its influence is being felt in other areas, and increasing, including the humanities. Its applications are transforming the way we interact with each other and with our environment. This book summarizes the current state of the art. (i.e. at the time of writing..!)

The Singularity is Near - Ray Kurzweil, Penguin Books, 2006, 672 pages, ISBN 978-0143037880. The celebrated inventor and futurist Ray Kurzweil is one of the best-known and controversial advocates for the role of machines in the future of humanity. In his latest, thrilling foray into the future, he envisions an event—the "singularity"—in which technological change becomes so rapid and so profound that our bodies and brains will merge with our machines.

The Singularity Is Near portrays what life will be like after this event—a human-machine civilization where our experiences shift from real reality to virtual reality and where our intelligence becomes non-biological and trillions of times more powerful than unaided human intelligence. In practical terms, this means that human aging and pollution will be reversed, world hunger will be solved, and our bodies and environment transformed by nanotechnology to overcome the limitations of biology, including death.

We will be able to create virtually any physical product just from information, resulting in radical wealth creation. In addition to outlining these fantastic changes, Kurzweil also considers their social and philosophical ramifications.

Moral Machines - Wendell Wallach and Colin Allen Oxford University Press, 288 pages, ISBN 978-0199737970. This important book draws attention to the need for more work on moral machines to begin now, before it is too late. Should intelligent computers be allowed to make moral decisions? Several present day repetitive type machines have already caused fatal accidents. Home and service robots will be with us soon, initially to assist the elderly.

Can a machine be a truly moral agent? What is the nature of Ethical Theory and the Least Resulting Disaster paradox? Can autonomous systems be ethical? Computers are already approving financial transactions, controlling electrical supplies, and driving trains. Soon, service robots will be taking care of the elderly in their homes, and military robots will have their own targeting and firing protocols.

Colin Allen and Wendell Wallach argue that as robots take on more and more responsibility, they must be programmed with moral decision-making abilities, for our own safety. Taking a fast paced tour through the latest thinking about philosophical ethics and artificial intelligence, the authors argue that even if full moral agency for machines is a long way off, it is already necessary to start building a kind of functional morality, in which artificial moral agents have some basic ethical sensitivity. But the standard ethical theories do not seem adequate. More socially engaged and engaging robots will be needed. As the authors show, the quest to build machines that are capable of telling right from wrong has begun. Moral Machines claims to be the first book to examine the challenge of building artificial moral agents, probing deeply into the nature of human decision making and ethics.

Robots Will Steal Your Job, But That's OK: How to Survive the Economic Collapse and Be Happy, 215 pages, 2014, ASIN B009R93JR6- "You are about to become obsolete. You think you are special, unique, and that whatever it is that you are doing is impossible to replace.

You are wrong. As we speak, millions of algorithms created by computer scientists are frantically running on servers all over the world, with one sole purpose: to do whatever humans can do, but better technological unemployment, one that is pervading modern society. But is that really the case? Or is it just a futuristic fantasy? What will become of us in the coming years, and what can we do to prevent a catastrophic collapse of society? This book explores the impact of technological advances on our lives, what it means to be happy, and provides suggestions on how to avoid a systemic collapse."

Robot Futures, The M.I.T. Press, 2013, 160 pages, ISBN-13: 978-0262018623. "With robots, we are inventing a new species that is part material and part digital. The ambition of modern robotics goes beyond copying humans, beyond the effort to make walking, talking creations that are indistinguishable from people. Future robots will have superhuman abilities in both the physical and digital realms. They will be embedded in our physical spaces, with the ability to go where we cannot, and will have minds of their own, thanks to artificial intelligence.

They will be fully connected to the digital world, far better at carrying out online tasks than we are." In Robot Futures, the roboticist Illah Reza Nourbakhsh considers how we will share our world with these creatures, and how our society could change as it incorporates a race of stronger, smarter beings."

Singularity Rising - James D. Miller, BenBella Books, 2012, 288 pages, ISBN 978-1936661657. Where will you be in a singular world? In Ray Kurzweil's New York Times bestseller, **The Singularity is Near,** the futurist and entrepreneur describes the singularity, a likely future utterly different from anything we can imagine. The singularity is triggered by the tremendous growth of human and computing intelligence that is an almost inevitable outcome of Moore's Law.

Since the book's publication, the coming of singularity is now eagerly anticipated by many of the leading thinkers in Silicon Valley, from PayPal mastermind Peter Thiel to Google co-founder Larry Page. The formation of the Singularity University, and the huge popularity of the singularity website, www.kurzweilai.com, speak to the importance of this intellectual movement. But what about the average person? How will the singularity affect our daily lives—our jobs, our families, and our wealth? This book focuses on the implications of a future society faced with an abundance of human and artificial intelligence. James D. Miller, an economics professor and popular speaker on the singularity, reveals how natural *selection has been increasing human intelligence over the past few thousand years and speculates on how intelligence enhancements will shape civilization over the next forty years.

Smarter Than Us: The Rise of Machine Intelligence, Stuart Armstrong, Machine Intelligence Research Institute, 2014, 64 pages, ASIN B00IB4N4KU. What happens when machines become smarter than humans? Forget lumbering Terminators. The power of an artificial intelligence (AI) comes from its intelligence, not physical strength and laser guns. Humans steer the future at present, not because we're the strongest or the fastest but because we're the smartest. When machines become smarter than humans, we'll be handing them the steering wheel. What promises—and perils—will these powerful machines present? Stuart Armstrong's new book navigates these questions with clarity and wit.

Can we instruct AIs to steer the future as we desire? What goals should we program into them? It turns out this question is difficult to answer! Philosophers have tried for thousands of years to define an ideal world, but there remains no consensus. The prospect of goal-driven, smarter-than-human AI gives moral philosophy a new urgency. The future could be filled with joy, art, compassion, and beings living worthwhile and wonderful lives— but only if we're able to precisely define what a "good" world is, and skilled enough to describe it perfectly to a computer program.

AIs, like computers, will do what we say—which is not necessarily what we mean. Such precision requires encoding the entire system of human values for an AI: explaining them to a mind that is alien to us, defining every ambiguous term, clarifying every edge case. Moreover, our values are fragile: in some cases, if we mis-define a single piece of the puzzle—say, consciousness—we end up with roughly 0% of the value we intended to reap, instead of 99% of the value.

Though an understanding of the problem is only just beginning to spread, researchers from fields ranging from philosophy to computer science to economics, are working together to conceive and test solutions. Are we up to the challenge? A mathematician by training, Armstrong is a Research Fellow at the Future of Humanity Institute (FHI) at Oxford University. His research focuses on formal decision theory, the risks and possibilities of AI, the long term potential for intelligent life (and the difficulties of predicting this), and anthropic (self-locating) probability. Armstrong wrote **Smarter Than Us** at the request of the

Machine Intelligence Research Institute, a non-profit organization studying the theoretical underpinnings of artificial super-intelligence.

Intelligent Systems for Engineers and Scientists, Third Edition, Adrian A. Hopgood. Published by CRC Press, 451 Pages, 180 Illustrations, ISBN 9781439821206.This very comprehensive book covers the full breadth of artificial intelligence techniques in a single accessible volume. It demonstrates how to apply artificial intelligence techniques with examples from engineering, science, and technology, and includes access to an AI toolkit with example programs that readers can download and test. It examines the principles of artificial intelligence and their application to engineering and science, as well as techniques for developing intelligent systems to solve practical problems. It covers the full spectrum of intelligent systems techniques, and incorporates knowledge-based systems, computational intelligence, and their hybrids.

Using clear and concise language this third edition features updates and improvements throughout all its chapters. It includes expanded and separated chapters on genetic algorithms and single-candidate optimization techniques, while the chapter on neural networks now covers spiking networks and a range of recurrent networks. The book also provides extended coverage of fuzzy logic, including type-2 and fuzzy control systems.

Example programs using rules and uncertainty are presented in an industry-standard format, so that you can run them yourself. The first part of the book describes key techniques of artificial intelligence—including rule-based systems, Bayesian updating, certainty theory, fuzzy logic (types 1 and 2), frames, objects, agents, symbolic learning, case-based reasoning, genetic algorithms, optimization algorithms, neural networks, hybrids, and the Lisp and Prolog languages. The second part describes a wide range of practical applications in interpretation and diagnosis, design and selection, planning, and control. The author provides sufficient detail to help you develop your own intelligent systems for real applications.

Our Final Invention - Artificial Intelligence and the End of the Human Era, Thomas Dunne Books, 2013, 336 pages, ISBN 978-0312622374. "Artificial Intelligence helps choose what books you buy, what movies you see, and even who you date. It puts the "smart" in your smartphone and soon it will drive your car. It makes most of the trades on Wall Street, and controls vital energy, water, and transportation infrastructure. But Artificial Intelligence can also threaten our existence.

In as little as a decade, AI could match and then surpass human intelligence. Corporations and government agencies are pouring billions into achieving AI's Holy Grail—human-level intelligence. Once AI has attained it (AGI), scientists argue, it will have survival drives much like our own. We may be forced to compete with a rival more cunning, more powerful, and more alien than we can imagine.

Through profiles of technical visionaries, industry watchdogs, and ground breaking AI systems, Our Final Invention explores the perils of the heedless pursuit of advanced AI. Until now, human intelligence has had no rival. Can we coexist with beings whose intelligence greatly exceeds ours, and will they want that?"

The Age of Intelligent Machines, The MIT Press, 1992, 579 pages, ISBN 978-0262610797 "Inventor and visionary computer scientist Raymond Kurzweil probes the past, present, and future of artificial intelligence, from its earliest philosophical and mathematical roots to tantalizing glimpses of 21st-century machines with superior intelligence and truly prodigious speed and memory. Generously illustrated and easily accessible to the non-specialist, this book provides the background needed for a full understanding of the enormous scientific potential represented by intelligent machines as well as their equally profound philosophic, economic, and social implications. Running alongside Kurzweil's historical and scientific narrative are 23 articles examining contemporary issues in artificial intelligence. Raymond Kurzweil is the founder and chairman of Kurzweil Applied Intelligence and the Kurzweil Reading Machine division of Xerox. He was the prime developer of the first print-to-speech reading machine for the blind and other significant advances in artificial intelligence technology.

Facing the Intelligence Explosion, Machine Intelligence Research Institute, eBook edition, 91 pages, ASIN: B00C7YOR5Q. "Sometime this century, machines will surpass human levels of intelligence and ability. This event—the "intelligence explosion"— will be the most important event in our history. Navigating it wisely will be the most important thing we can ever do. Luminaries from Alan Turing and I. J. Good to Bill Joy and Stephen Hawking have warned us about this. Why do I think Hawking and company are right, and what can we do about it? Facing the Intelligence Explosion is my attempt to answer these questions." – Luke Muehlhauser.

The Human Race to the Future, What Could Happen - and what to do about it, 2014, 394 pages, ISBN 978-1494712112. "Substantive yet imaginative, readable, occasionally humorous, and science oriented, this book proposes future scenarios spanning from the current century to nearly eternity. Most chapters offer a concluding section with recommendations and often, agree or disagree, with the author's occasionally inimitable opinions. Some of the recommended actions can be done by individuals, others by nations or other groups, and still others by the entire world."

Chapters cover and include, Keyboards yesterday and Mind Reading tomorrow. The future Singularity. Will Artificial Intelligence threaten our civilization?

THE AVOGADO CORP. by William Hertling, 300 pages, Liquididea Press. *This book is a perfect example of speculative 'near future' fiction, that you must read. It describes how an Artificial Super Intelligent software program might emerge accidentally. It is a truly brilliant analysis of what could happen when an intelligent algorithm is uploaded into a sufficiently powerful computer network. This clever author knows his subject forwards and backwards, and how to blend it into a plausible, exciting tale. Don't miss this one. I'm very glad I didn't.* - Review by Tony Thorne MBE.

The Avogadro Corp Trilogy The Singularity Is Closer Than It Appears, A.I. Apocalypse, and The Last Firewall. Hertling's near-term science-fiction novels about the realistic ways strong AI might emerge, have been called "frighteningly plausible", "tremendous", and "a must read". He's been influenced by writers such as William Gibson, Charles Stross, Cory Doctorow, and Walter Jon Williams.

William Hertling was born in Brooklyn, New York. He grew up a digital native in the early days of bulletin board systems. His first experiences with net culture occurred when he wired seven phone lines into the back of his Apple IIe, creating an online chat system. He currently resides in Portland, Oregon. By day he works on web and social media for HP. Follow him on twitter at @hertling or visit his blog williamhertling.com.

A.I. APOCALYPSE by William Hertling, 264 pages, ISBN: 0984755748, Liquididea Press; ASIN: B007FZVI2M. The saga continues at an amazing pace. Thoroughly recommended for a non-stop, mind gripping, breathtaking read.

THE LAST FIREWALL, William Hertling, 323 pages, ASIN B00EEIGHDI. *An entertaining read, absolutely packed with continuous action. Highly relevant, despite the occasional bad language lapse... but perhaps I'm just a prim and proper old-timer.*

... and finally, this recent extract to think about...

"The very ease of digital recording and transmitting— the breakthrough that permits software and data to be, in effect, immortal—removes robots from the world of the vulnerable (at least robots of the usually imagined sorts, with digital software and memories). If this isn't obvious, think about how human morality would be affected if we could make "backups" of people every week, say. Diving headfirst on Saturday off a high bridge without benefit of a bungee cord would be a rush that you wouldn't remember when your Friday night backup was put online Sunday morning, but you could enjoy the videotape of your apparent demise thereafter.

So what we are creating are not—should not be—conscious, humanoid agents but an entirely new sort of entity, rather like oracles, with no conscience, no fear of death, no distracting loves and hates, no personality (but all sorts of foibles and quirks that would no doubt be identified as the "personality" of the system): boxes of truths (if we're lucky) almost certainly contaminated with a scattering of falsehoods.

It will be hard enough learning to live with them without distracting ourselves with fantasies about the Singularity in which these AIs will enslave us, literally. The human use of human beings will soon be changed—once again—forever, but we can take the tiller and steer between some of the hazards if we take responsibility for our trajectory.

From "What Can We Do?" by Daniel C. Dennett. Adapted from Possible Minds: Twenty-Five Ways of Looking at AI, edited by John Brockman, published by Penguin Press, an imprint of Penguin Publishing Group, a division of Penguin Random House LLC. Copyright © 2019 by John Brockman. Daniel C. Dennett is the Austin B. Fletcher professor of philosophy and codirector of the Center for Cognitive Studies at Tufts University.

Section 5. About the 'author/compiler'of this book, and some of his other work.

Tony Thorne MBE, M.I.Mech.E (rtd), M.I.Nuc.E. (rtd) qualified as a Chartered Design Engineer, in London, UK, and then built up a research and development (nuclear and medical applications) company in Kent, England, specializing initially in 3000 degrees Celsius, (laboratory and industrial) nuclear grade graphitizing furnaces, and then subsequently, units for carbon fiber processing.

Other developments included vacuum super-insulated pipelines for liquefied nitrogen, oxygen and helium gases, nuclear and industrial protection equipment, miniature thermocouples and finally instrumentation for low temperature surgery of the human eye. The latter and its worldwide commercial success led to him being awarded an MBE by the Queen in 1968. He was also awarded patents for some of these developments.

He later 'brain-drained' to the American Dynatech Corporation and set up their medical products, international operations, resulting in him being appointed as CEO European Operations, and a Vice President of the parent Dynatech Corporation, in Boston, USA. He was also awarded an American patent for developments in laboratory micro-titration equipment, used for microbiological assay studies. He stayed with Dynatech for sixteen years, until he retired to start up his own computer training operation, later specializing in AI software able to generate software code for business and graphics programs and animated (large screen) computer graphics programs set to music.

Later, he moved the operation to Somerset in England, and created software for Quarry Laboratory Quality Control systems, Hospital Financial Analysis programs and many other applications, including private in-house Lottery Systems.

Now 'retired', he writes near-future speculative fiction, mostly of the science and fantasy genre and has published fourteen collections of short stories, including the awards winning TENERIFE TALL TALES trilogy, and MACABRE TALES. These were soon followed by ROBOTS INCLUDED, and TALL SF TALES FOR TEENAGERS, all available, as eBooks and paperbacks.

His near future SciFi Espionage novels, the POINTS OF VIEW series now comprise four volumes, published by Etcetera Press.

Tony Thorne MBE's SF, Macabre and other, publications so far include**Speculative Fiction**: TENERIFE – Tall Tales with a Twist (Whortleberry Press, USA), POINTS OF VIEW (Eternal Press, USA), TIME TRAVEL TROUBLE, FUTURE REASSURED, FUTURE UNCERTAIN, ROBOTS INCLUDED, TALL SF TALES FOR TEENAGERS, MORE TALL TENERIFE TALES, EVEN MORE TENERIFE TALES, BILL's HELICAL BLUES, THE BEST OF THE TENERIFE TALL TALES, SPECULATIVE TALES, and MACABRE TALES (Etcetera Press)

Satire/Semi-Autobiographical: BEST SELLING AUTHOR PLAN, HOW TO BE A TOP EXECUTIVE, THE QUALITY OFLIFE, THE JUNIOR PHILOSOPHICAL
SOCIETY, INSIDE INFORMATION (Etcetera Press Publications)

Poetry: SECOND OPINION [The Poets Yearbook], SHOWCASE –The Guernsey Poets (Editor/contributor), and the websites AUTHOR'S DEN, SCIFINITY, and ABANDONED TOWERS MAGAZINE.

Plus Stories, Poems & Articles in various magazines, anthologies and websites, including PLANET, NEBULA, BSFM, CATS AROUND THE CHRISTMAS TREE, SANTA'S GIVINGS & MISGIVINGS, STRANGE MYSTERIES 1 to 5, Whortleberry Press, USA, Anthologies); CREATURES OF GLASS & LIGHT

(The EuroCon2007 anthology), AXXON (Argentina SciFi Magazine), SPICK, ORBIT, ART & PROSE MAGAZINE, ALIEN SKIN MAGAZINE, LET'S START AGAIN (USA anthology, ASTONISHING STORIES (American magazine).

Some Reviews of THE SINGULARITY IS COMING - The Artificial Intelligence Explosion - eBook and paperback.

"Dear Tony, Your book Singularity raises many degrees of awareness about the dangers of creating minds superior to ours. As you point out, it would be very soon that we would lose control of an electronic identity and it would be free to engineer its own supremacy.

Logically, giving this some thought, any intelligent race of beings would have gone down this path long before us. If the Drake equation does indeed point to the possible number of civilizations scattered amongst the stars being a vast number, the question of where are they, the Fermi paradox is now answered! If they have invested their time and efforts producing A.I. then the 'beings' that they created would have no use for organics and would spread throughout the galaxies seeking 'dry' planets to expand into rather than water planets like ours. Any electronic intelligence once aware of us would just wait until our creation dominates this world and then it might desire contact. Hopefully learning the futility of warfare amongst its own kind as the universe is after all almost infinite and there is plenty of room for something that does not need to breathe! Thanks for the sleepless nights! "

An Outstanding Experience! ***** This book outlines the concept of a technological singularity. It explains how a very dangerous one will arrive relatively soon and I am amazed at how well this subject has been researched and thought out. The storyline examples are suspenseful and full of interesting twists and turns. They, and the whole subject, never become dull or boring. This is one of those rare 'fact to fiction' books where I was hooked into reading it through from start to finish.

I always enjoy a well-researched book, and the author has obviously done his research on this one. Ever since I read William Gibson's Necromancer, I have had a fascination for computer AIs and this one covers the areas I have thought about and goes beyond. I often wonder what type of AI will appear first the benign or not so benign and the survivability of the human nature is linked to, and the book is easy to understand for the non-techies out there.

The second part comprises linked sci-fi tales, an impressive idea, I like my sci-fi researched and I know this author has done his homework. - By Seonicked (Reviewer pen name)

Try something different from this award winning, speculative fiction, writer of near future tales. It anticipates some of the amazing events we may well experience in less than 10 years...! One thing is certain, the last 75% of this century will be like nothing we've experienced before. Nanotechnology and Artificial General Intelligence will see to that. Many experts think we have a minimal chance of surviving the advent of super intelligent machines. - By Emma G.- Website pages.

Fascinating Collection!!! - November 7, 2014 - **Robots Included** is a fascinating collection ranging in different points of view and perspectives of a future with more advanced robots and artificial intelligence. I appreciated the introduction with the background references to the different definitions of robotic words used (i.e. cyborg versus android) as well as Asimov's Three Laws of Robotics. T he writing itself is innovative and thoughtful. The language is clear, which I appreciate in science fiction when the advanced concepts can be confusing enough. The collection of stories all follow a theme of a possible future, be it near or far, that is impacted by the advancement of robotics. The stories can easily be read one at a time, but curiosity may cause the desire to read on until human needs like sleep and food require one to put the book down. These are not "cookie cutter" stories, so you will not read the same kind of story twice. Examples range from robots as a child's toy to protectors of an alien civilization.

I would recommend this book to teens who are developing an interest in science fiction or adults who are open to expanding their minds to future possibilities.

***** **Review** October 26, 2014 - Mind-gripping! **RobotsIncluded** is worth every penny. Its value and integrity exceeded my anticipation. The robots in this book are mind-gripping creations. I find it difficult to adequately describe the overall composition and operation of the author's robots. They have unexpected ways of carrying out their duties according to how they are programmed. This riveting book unveils the various types of AI robots that could soon exist and the resulting stories are astounding. Machines at present are totally dependent on humans to do the necessary adjustments and fixes—but for how long?

TENERIFE TALL TALES Vol. 1, by Tony Thorne MBE. "Tenerife holds a special place in the heart of this author and his stories detail the local culture and ecology in a most interesting way. His tales journey from a micro-Jurassic Park with containment issues, past intelligent dolphins, seals, lizards and genetically engineered chimpanzees, around magic belts and jumping F1 racing cars, through black holes and finally with a three-part look at a dismal but hopeful future for our planet.

Although I enjoyed all of them, the following three really stood out: "Hologhosts", a story about five friends who imagine themselves amateur "ghost-hunters", and having discovered that history can be stored in certain kinds of rock and emitted as a holographic image, these friends set up camp alongside Teide, a volcano in Tenerife. "Baggage Included", is a funny little story about a lost alien, an interstellar transport system and the front half of a local TITSA commuter bus, and in "Retirement Plan" where Thorne ventures into the macabre, the moral of this story applies to us all, however close to retirement we may be." – Kelly Jensen, reviewer, SF Newsletter.

"With all the fervor and creativity an author can muster, Tony Thorne MBE has put together a remarkable collection of short stories, inspired by the island he loves and the genre of shadowy imagination that he is so very good at capturing. Thorne's stories are meticulously laced with a perfect concoction of urban legend, scientific precision, and his own edgy yarns to be almost believable, yet far beyond the imaginations of what we know as reality. Anyone who enjoys good fiction with a tinge of daunting possibility should consider **Tenerife Tales** a must for their reading list." – Autumn Conley, American author of two novels, many articles, reviews and short fiction items for many USA magazines and newspapers.

"**Despite a solid scientific background,** Tony is cynical and distrustful of what the future may hold and has drawn on that to produce short stories, all with Tenerife connections, to startle, amaze and warn us all. Mix in a few anecdotal observations and you have a guidebook to the bizarre and downright strange, to lead you on flights of fancy. You shouldn't take it too seriously as Tony describes his form of fiction as speculative. His easy style will carry you through his tales with belief suspended until you reach the end and think, well that just couldn't happen, and of course it couldn't, could it?" – Colin Kirby, freelance journalist, who writes for The Tenerife Sun, Think Spain Today and The Daily Record.

Operation Mayberry - Outstanding! 5 Stars! Review of **Points of View Volume 1**, by Tony Thorne MBE. "I can't say enough about this book, 'Wow' is one word, but seriously I can really see it as a movie. I started to read it and was hooked. I am not much of a Sci-Fi reader but after reading this, I welcome them. Points of View is about a young Englishman, Horace Mayberry, who became blind at early age, with no help from the medical resources available to him. Resigned to his fate, he enjoys dreams about being a government superspy. They are his only way of seeing the world, being unable to see through his own eyes.

He accepts a free offer to replace his eyes, using new developments in nanotechnology. However, there is catch, he must become assistant to a real government secret agent, Captain Aubrey Jackson. The idea of his new job is exciting because he can now live his dreams for real, even though he just wants to see the world that he missed when he was blind. However, he is soon thrown in a world of power and denials, and becomes a man wanted for his super-human eyes, to be used for both good and evil.

This is a thrilling ride of adventure, mystery, action, humor, and even some romance. I was sucked into this book. I loved it. I couldn't read the pages fast enough to know what was to happen next. I can see this book as a Movie, because I could see the action happening as I was reading it. I love books that open up the mind to your own private showing of the story. I highly recommend this book to all book lovers. It has a great storyline that will not disappoint you. This is a, don't miss, phenomenal 5 stars, tale of twists and turns and with some humor. I look forward for more books from Tony Thorne MBE." – Ana Torres, USA, Reviewer.

POINTS OF VIEW (Volume 1) By Tony Thorne, MBE - Review by Rhetta Akamatsu, the author of Haunted Marietta and other books related to history and the paranormal, as well as an avid science fiction fan and reviewer. "What if you were blind and suddenly the government offered you a way to see again, with Nanotronic BioVision, allowing you to see the world, literally, in a whole new way? In return, you are to use your new points of view to serve your country, but other people want to take advantage of your unique abilities as well. That is precisely the position faced by Horace Mayberry, the central character of Tony Thorne's new series of novels, Points of View. Thorne, a design engineer who earned an MBE for his expertise, uses his knowledge of cutting edge technology to fashion a fascinating technological thriller, just barely in the future. With lots of authentic detail, Points of View is believable and absorbing. I highly recommend this book to any fan of speculative fiction."

"**Macabre Tales** is a collection of hypothetical, scientific and theoretical short stories. The author has observed some daily life events and written speculative stories on that. Each story is entertaining and full of suspense. e.g. In Dead Ringer, a dead body takes revenge by a phone call. Slimming Plan story is full of suspense, thrills and depicts a murder case. Each story is short and to the point. The author has proved his out-of-the box thinking skills. His imagination coupled with scientific knowledge has made this book totally different.

The stories are so interesting, and the characters are so vivid that I could not stop my temptation to complete a story after starting it, and once one story was finished, I was tempted to read the next. I really enjoyed this book. It is one of my favorite reads. Each story fits in the context and really does a perfect job of hitting every angle you can think of. This book is a must read for people interested in science fiction, speculative and hypothetical reading, but I would recommend this book to everyone." - Naina John,
Pro-Reviewer, USA.

Macabre Tales - by Tony Thorne MBE, who takes Readers into Strange, sometimes Weird, Places! "If you stop to think about the everyday things that happen in your life, and then turn a, scientific, but definitely weird glare on that activity, you would begin to understand the type of stories written by Tony Thorne.

For instance, his parting tale describes how a man finds his wife frozen in fright, holding a knife, and standing near their washing machine. The man had already been frightened by a strange creature on the way home, so immediately he wondered if another one was in the machine. After all, something was causing the glow and the strange wailing coming from inside it... I loved the ending of this one!

A famous American SciFi author, Harry Harrison, described Thorne as having the mind of a scientist, and the soul of an artist...[with] a brilliant flash of black humor. Indeed, many of the stories you will read deal with the potential misuse of science. Consider the cell phone. Do you carry yours everywhere you go? Might you even take it to the grave with you? After all, you might want to, say, call your unfaithful spouse and share a little...
Have you heard about nanotechnology? It is the future for medical surgical procedures according to some individuals... Thorne poses the question: What if one of those 'intelligent' nanophytes decides to take up residence in...your head? In these days of hacking, identity and information theft on the computer, Thorne postulates that we might have to fight fire with fire, as the old saying goes...or, rather, fight with anything that is in range of a computer screen--even a fly..! A futuristic thought... individuals may become trained to work at Euthanasia, Inc. You'd probably never be out of work, right? Don't count on it...

We are all warned about having radiation therapy except in short bursts; it's dangerous so even workers don't stay around when the machine is on. But if a mouse should crawl way up into the internal parts of that equipment... Or consider the weight-loss clinic which hosts free operations to those qualified... but with certain guarantees.

I found Tony Thorne more droll than truly gruesome. His concepts are clever, his writing sharp and quickly to the point and he closes with a quiet flourish. He delves into personal areas, our daily lives and asks you to consider the alternatives that are conceivably possible, if only... But then, again, there really might be a bit of truth just waiting to be scientifically researched and created!

Explore our potential future with Tony Thorne, and recognize that some tales are indeed possible right now! ***** Highly recommended for a true exploration of the macabre."
– G A Bixler Reviews

All the Author's books, EBook versions and Paperbacks, are available from most outlets.

Finally, if you have found this book to be interesting, or even if otherwise, the author would appreciate your comments, via the Contact Me link on his website: www.tonythornembe.com. Even better, if you purchased this book, a short review or whatever, on your supplier's website would be gratefully received. That's the only effective way authors have of receiving essential publicity for anything nowadays.

The author's website is: www.tonythornembe.com

An Etcetera Press Publication

A Chinese language version of this book was published in June 2016 by the Chinese IT Books Division, PTPress in Beijing, in paperback and eBook editions.